THE EINSTEIN ALMANAC

The Einstein Almanac

Alice Calaprice

The Johns Hopkins University Press
Baltimore and London

 Published 2005
Printed in the United States of America on acid-free paper
9 8 7 6 5 4 3 2

The Johns Hopkins University Press
2715 North Charles Street
Baltimore, Maryland 21218-4363
www.press.jhu.edu

Library of Congress Cataloging-in-Publication Data

Calaprice, Alice.
The Einstein almanac / Alice Calaprice.
p. cm.
Includes bibliographical references and index.
ISBN 0-8018-8021-1 (acid-free paper)
1. Einstein, Albert, 1879–1955. 2. Physicists—Biography.
3. Relativity (Physics)—History. 4. Physics—History—20th century. 5. Einstein, Albert, 1879–1955—Bibliography. I. Title.
QC16.E5C35 2004
530'.092—dc22 2004009048

A catalog record for this book is available from the British Library.

CONTENTS

PREFACE

The past exists only insofar as it is present in the records of today. And what those records are is determined by what questions we ask. There is no other history than that.—John Wheeler, 1982

When looking at Albert Einstein's entire list of publications, one can't help but feel awe at the breadth and scope of his talents, interests, and involvements and, ultimately, his influence. Surely, no other scientist has been able to match such a legacy in the twentieth century.

This book presents a concise summary of Einstein's life and of the times in which he lived, stressing the major landmarks in physics and, within that context, giving a large sampling of his work. A reader can discover the influences and circumstances in Einstein's life that may have led to his immense contributions. The accident of the historical time into which Einstein was born allowed him to mature into adulthood during an era of exciting cultural and political changes. Other factors that influenced his life were his family background and values and his ethnicity; the people he serendipitously encountered; his unique genetic makeup and personality; and, indeed, the scientists who came before him and made his discoveries possible.

In the pages that follow, you'll find most of Einstein's publications, speeches, and contributions to his community, the last of which he felt were paramount in anyone's life. Albert Einstein began his long list of publications with a scientific paper he had finished writing at the age of twenty-one. It was published in the German journal *Annalen der Physik* in March 1901. The book closes with a nonscientific article for *Common Cause* in 1955. Such a beginning and such an end seem appropriate, for Einstein did his best science during the first half of his life—and some even say only in the decades between 1905 and 1925, when he was relatively young. His interests broadened as the events of the twentieth century paved the way for his increasing commitment to international political and social issues.

In between these two papers we find an assortment of publications from one of the most multifaceted individuals of our time. Einstein's writings reveal his wide-ranging interests and viewpoints. Fortunately for us, he was a man of many words and convictions, fearless and sometimes ruthless in expressing his ideas and opinions. He had something to say on everything from relativity and quantum theory to a "theory of living," from German nationalism to liberal Judaism, from war to peace, from atheism to "cosmic religion," from lynching to the death penalty.

For this book I have chosen only a representative selection from his huge résumé of more than six hundred diverse publications, concentrating mostly on his scientific and humanitarian writings. In addition, I have included a small sampling of the interviews he gave and the eulogies and tributes he wrote. He also participated in radio broadcasts and wrote book reviews, patent opinions, and newspaper articles. Somehow, he also found time to write thousands of letters and a variety of statements, aphorisms, and speeches that were not published but were later collected and printed in compilations such as *Ideas and Opinions* half a century ago and more fully in the ongoing series, *The Collected Papers of Albert Einstein.* Some of these gems can be found in my *Quotable Einstein* books.

To put these works into context, I am presenting them chronologically with descriptions of concurrent events in Einstein's personal life, the world in general, and the realm of physical science. I summarized the content of each paper whenever I was able to find information on it or had access to the book or article itself.

I thank the efficient and kind Osik Moses of the Einstein Papers Project for taking the time to send me some articles from the *Readex* bibliography, which saved me hours of library work. I gratefully acknowledge the rigorous editing of the manuscript by Linda Forlifer of the Johns Hopkins University Press. I am also indebted to the anonymous reader of the manuscript for valuable suggestions, most of which I have tried to integrate into the text.

The idea for this project came from Trevor Lipscombe, editor in chief at the Johns Hopkins University Press. I owe much to Trevor: he sponsored my first Einstein books, the popular *The Quotable Einstein,* followed by the *Expanded Quotable Einstein,* while he was physics editor at Princeton University Press, paving the way for my subsequent Einstein publications, some still in progress at this writing. I dedicate this book to him and his family. I thank him for giving me the chance to put this almanac together as what I hope is an interesting exposition of the amazing life of an extraordinary man.

A BRIEF EINSTEIN TIMELINE FOR THE YEARS 1879–1900

1879 March 14: Albert Einstein was born in Ulm, Germany, to Jewish parents, Hermann and Pauline Koch Einstein. Scottish mathematical physicist James Clerk Maxwell, who unified electricity and magnetism into a single, coherent electromagnetic theory and who was a pioneer in the realm of the kinetic theory of gases, died this same year. Thomas Edison and Joseph W. Swan independently devised the first usable electric lights—the former in Menlo Park, New Jersey; the latter in Newcastle, England.

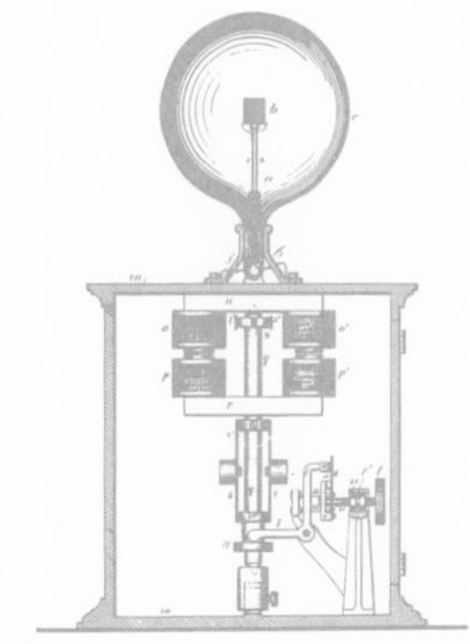

Edison patents an electric lamp.

Former slave Frederick Douglass is an influential figure in America around the time of Einstein's birth.

1880 The Einstein family moved to Munich. In France, Louis Pasteur performed the first inoculation, on chickens against fowl cholera.

1881 Einstein's sister, Maja, was born. Thinking that his parents were presenting him with a new toy—perhaps a pull toy he had wanted—the two-year-old asked his parents where the wheels were.

1884 Young Einstein's father showed him a compass, which fascinated the five-year-old, making him aware of forces that can't be seen.

1885 In the fall, Einstein began his education at a Catholic neighborhood school, the only Jew in his class. He received Jewish religious instruction at home and became curious about religion, a subject that would fascinate him throughout his life; he also began violin lessons. Louis Pasteur devised a rabies vaccine, and Darwin's half cousin, Sir Francis Galton, proved that each person's fingerprints are unique. Danish physicist Niels Bohr, with whom Einstein would hotly debate quantum theory, was born.

1886 Einstein discontinued his violin lessons, but the seven-year-old continued to practice the instrument and also taught himself to play the piano. In the United States, Edison and Swan collaborated to produce "Ediswan" electrical lamps.

1887 Ernst Mach developed the supersonic scale. The American physicists Albert Michelson and Edward Morley attempted to measure the velocity of the earth through the "ether." Using an interferometer, an instrument designed to produce optical interference fringes (a series of faint, irregular, roughly parallel lines of different colors sometimes seen in clear insulating glass), they expected to see a shift in the fringes formed when the instrument was rotated through 90 degrees. This would show that the speed of light measured in the direction of the earth's rotation is not identical to its speed at right angles to this direction. But they observed no shift. Later theorizing on this experiment would contribute to Einstein's work: An explanation by George F. Fitzgerald in 1892 and independently by Hendrick A. Lorentz in 1895 showed that the "null" result could be explained by the shortening of an object along the direction of its motion as its speed approaches the speed of light, as measured by an observer at rest with respect to the object. This "Lorentz-Fitzgerald contraction" was an important step in the mathematical formulation of Einstein's special theory of relativity and relegated the ether to the history books.

In the aftermath of Ireland's potato famine (1845–50), peasants seize the crops of evicted tenants in 1886.

1888 German emperor Wilhelm I died in March. He was succeeded by his son Frederick III, who died in June and was then succeeded by his son Kaiser Wilhelm II. In the United States, Croatian American engineer Nikola Tesla, who invented the rotating magnetic field, the basis for most alternating-current machinery, built an electric motor, manufactured by Westinghouse. George Eastman perfected the Kodak box camera. Heinrich Hertz demonstrated the wavelike properties of electromagnetic radiation, decisively confirming Maxwell's theory of electromagnetism. His experiment led to the development of the wireless telegraph and radio.

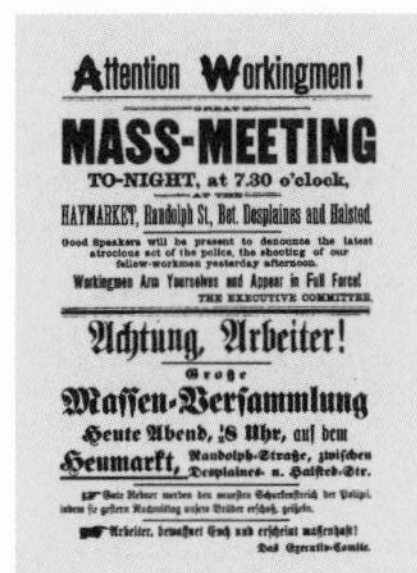

Civil unrest follows the police shooting of unarmed laborers in Chicago (May 1886).

1889 At the age of ten, Einstein's interest in physics, mathematics, and philosophy began when a family friend, a university student who recognized Albert's intelligence and curiosity, introduced him to these subjects through popular scientific and technical books. In April, an Austrian couple welcomed their newborn son, Adolf Hitler.

1890 Eleven-year-old Albert was able to prove the Pythagorean theorem, and he enjoyed working out difficult problems and puzzles. An influenza pandemic, originating in Asia, swept the globe, killing tens of thousands.

1891 Pursuing readings on his own, Albert was now teaching himself higher mathematics and calculus. In Italy, a young scientist, Guglielmo Marconi, began experimenting with wireless telegraphy; his apparatus was based on the ideas of German physicist Heinrich Hertz, but he improved the design by grounding the transmitter and receiver and found that an insulated antenna enabled him to increase the distance of transmission.

1892 While Einstein was becoming a good violinist and continued to read science books voraciously, the first automatic telephone switchboard was developed. German engineer Rudolf Diesel, living in France, patented the internal combustion engine that bears his name. In 1913 Diesel would vanish from a steamer bound for London, and his body would be found ten days later, washed ashore.

1893 American Henry Ford built a gas engine, and German engineer Karl Benz built the world's first inexpensive, mass-produced four-wheel car, the Velo, beginning a new era in personal transportation. Though the automobile became indispensable to many people over the next half-century, Einstein never learned to drive, perhaps because others were willing to do it for him. The first photoelectric cell was developed by Julius Elster and Hans Geitel.

1894 Einstein's family moved to Italy; fifteen-year-old Albert stayed in Munich to finish school but was unhappy and quit, joining his family at the end of the year. His teacher had told him that he would never amount to anything

Deputies try to move an engine and car as railroad strikes cripple America in 1894.

and that his presence undermined the whole class's respect for the teacher.

1895 Einstein attempted to enroll in the Federal Polytechnical School (the "Poly," later the ETH, or Federal Institute of Technology) in Zurich two years early, but he failed the nonscientific part of the entrance exam and was urged to spend another year in secondary school. He did so in the village of Aargau, where he lived in the home of his headmaster. Later he would have fond memories of this year of his life. The Lumière brothers, Louis and Auguste of France, invented a movie camera (cinematograph), a twelve-pound machine that effectively served as a camera, projector, and printer all at once, and the first public film was shown in Paris. Wilhelm Roentgen discovered x-rays, and King C. Gillette invented the safety razor.

1896 At the age of seventeen, Einstein relinquished his German citizenship, with his father's consent, because he detested the country's obsession with regimentation in most aspects of life, and he remained stateless for the next five years. He entered the Poly in October. Upon the death of Swedish inventor and industrialist Alfred Nobel, five annual Nobel Prizes were established—in physics, physiology and medicine, chemistry, literature, and peace, to be awarded beginning in 1901. Antoine Henri Becquerel discovered natural radioactivity, and Guglielmo Marconi, at the age of twenty-two, patented a successful system of radio telemetry. Ludwig Boltzmann provided what he felt was a reasonable basis for statistical mechanics, expressing entropy in terms of probability theory. Sigmund Freud, who would correspond with Einstein some thirty years later, used the word *psychoanalysis* for the first time in his paper, "The Etiology of Hysteria." Zionism, the national movement to return Jews to their homeland in Israel, was founded as a response to anti-Semitism in western Europe and to violent persecution of Jews in eastern Europe. Einstein would later take an interest in Zionism and in the establishment of the state of Israel.

1897 In college, Einstein preferred to spend time in the physics laboratory, immersing himself in self-study, and borrowed lecture notes from classmates. J. J. Thomson discovered electrons, the first truly indivisible particles.

1898 In France, Pierre and Marie Curie discovered radium and polonium. Graf von Zeppelin, a German army officer, built his airship (the first "blimp"). The first photographs using artificial light were taken, and the first patent was given for a magnetic sound recording, the "telegraphone." Guglielmo Marconi successfully transmitted wireless signals across the English Channel. While working in Canada, New Zealander Ernest Rutherford reported on alpha and beta particles in uranium radiation and indicated some of their properties. The former were easily absorbed by matter; the latter were more penetrating.

USS *Maine* in Havana Harbor, sunk at the beginning of the Spanish-American War

British troops pose before engaging in vicious battles during the Boer War.

1899 Einstein applied for Swiss citizenship and spent his summer vacation with his mother and sister in Switzerland.

1900 At twenty-one, Einstein graduated from the Federal Polytechnical School and began a job search in Europe. At the same time, he began to work on problems in theoretical physics. Sigmund Freud published *The Interpretation of Dreams,* and F. E. Dorn discovered radar.

SELECTED PUBLISHED PAPERS & COMMENTARY, 1901–1955

The continent of Europe at the turn of the century

1901

At the turn of the century, Einstein had just graduated from the Swiss Federal Polytechnical School, the "Poly," in Zurich and was intensively looking for work. His carefree student days were over. Though he had been a good student, he had not relished attending classes, preferring to study on his own—using source books such as those by Hermann von Helmholtz, Heinrich Hertz, and Ludwig Boltzmann—and to discuss the topics of the day with his close circle of freethinking friends. Consequently, he now found it difficult to get favorable recommendations from his teachers, who had found him somewhat arrogant, sharp-tongued, and cocky. The Poly had already turned down his application to become an assistant teacher for the fall semester.

In 1896, Einstein, with his father Hermann's consent, had relinquished his German citizenship and moved to Switzerland to go to school. Einstein had long harbored an acute distaste for the authoritarian and military attitudes of strict regimentation and discipline popularized under German chancellor Otto von Bismarck and Kaiser Wilhelm II. Now, five years later, he became a citizen of the more tolerant nation of Switzerland, "the most beautiful corner on Earth I know." Ironically, three weeks later he was required to register for the military, an obligation he had sought to avoid in Germany. Fortunately for Einstein, however, he was declared unfit for military service due to flat feet, excessive foot perspiration (which may have led to his dislike of socks), and varicose veins. Instead of serving in the armed forces, he was required to make annual military tax payments. Einstein's sister, Maja, maintained that her brother wanted to become a Swiss citizen because he admired the nation's political system; however, he may also have wanted citizenship for the more utilitarian reason of becoming eligible for civil service positions, including teaching.

During his early college years, Einstein had become romantically involved with fellow physics student Mileva

Marić, a shy, bright, and intense young Serbian woman. Her noticeable limp, orthopedic boot, and moody disposition did not deter his feelings for her. Mileva was the daughter of an affluent Serbian landowner and judge in Novi Sad, a town in what was at that time a part of Hungary. She was four years older than Albert. At age twenty-two, after several years of courtship, he stated his intention to marry Mileva over his family's strong objections, especially his mother's. Mileva wrote to a friend in late 1901 that Mrs. Einstein "seems to have set as her life's goal to embitter as much as possible not only my life but also that of her son ... I would not have thought it possible that there could exist such heartless and outright wicked people!"

In the spring, Albert continued to apply for positions as an assistant to Swiss, German, and Dutch physicists but to no avail. With no other prospects in sight, in May 1901 he accepted a two-month position as a substitute teacher at the Technical School in the Swiss town of Winterthur, as he continued to be turned down for similar positions at other schools. For the fall semester, he accepted a temporary tutoring position at a private school in Schaffhausen and began work on a doctoral dissertation on the molecular forces in gases. He submitted this thesis to the University of Zurich in November (at that time the Poly had no Ph.D. program) and waited for a decision.

Adding to his anxieties, Mileva Marić had become pregnant with their child. Einstein, who at first seemed unperturbed by the impending event, realized soon enough that he was bereft of any means of support and hastily applied for a position at the Swiss Patent Office in Bern through arrangements made by the father of a school friend, Marcel Grossmann. Mileva had failed her final exams at the Poly for a second time and dropped out of school, returning home to Hungary to await the birth of her child. Einstein packed up his meager belongings and moved to Bern in optimistic anticipation of an offer of work from the Patent Office.

The city of Bern when Einstein lived there

WHILE THE YOUNG EINSTEIN was contemplating his bad luck that year, Italian inventor Guglielmo Marconi, on December 12, transmitted the first transatlantic wireless signals from Poldhu in Cornwall, England, to Saint John's in Newfoundland, Canada, a distance of twelve hundred miles. His system, radiating signals at about 850 kHz, proved that wireless waves were not affected by the curvature of the earth. The great Italian physicist Enrico Fermi, who would later develop the first nuclear pile at the University of Chicago, was born, and German physicist Max Planck devised quantum theory, in which he introduced the concept that energy existed in discrete units, called *quanta*.

The Nobel Prizes were awarded for the first time this year. The first Nobel Prize in physics was awarded to Germany's Wilhelm Roentgen in recognition of his discovery of the radiation subsequently named after him (also known as x-rays). The chemistry prize went to Jacobus van't Hoff of the Netherlands for discovering the laws of chemical dynamics and of the osmotic pressure in solutions.

1. **"Conclusions Drawn from the Phenomena of Capillarity" (Folgerungen aus den Capillaritätserscheinungen).** ***Annalen der Physik*, ser. 4 (1901): 513–523.**

Einstein dated his first publication December 13, 1900, though it was not published until the following March. Using both thermodynamic and molecular-theoretical methods, he examined the nature of intermolecular forces in the specific phenomena of capillarity in neutral liquids.

1902

Einstein's first child, daughter Lieserl, was born out of wedlock (probably in January) to Mileva, who had returned to her parents' home in Novi Sad to give birth. Einstein was not present for the birth, nor did he ever see his daughter. Indeed, Lieserl's story is shrouded in mystery. Her parents presumably gave her up for adoption or asked family or friends to care for her, and she may have died of scarlet fever around the age of two. Except for mentions of Lieserl in correspondence between Albert and Mileva before her birth and until the age of one and a half, investigators have been unable to find any record of her—no birth or death certificate—as if the baby girl had never existed. It is possible that public knowledge of the illegitimacy might have cost Einstein the civil service job he was seeking.

Probably on the advice of his thesis supervisor, Alfred Kleiner, Einstein withdrew his doctoral dissertation from the University of Zurich in February because the faculty considered it controversial and his ideas could not be proven experimentally. Discouraged and sullen about this latest setback, he accepted a provisional position at the Patent Office in Bern. He began his civil service career in June as a "technical expert third class" examining electrical patents. It was in Bern, where he continued to live for seven years, that Einstein, finally freed from economic worries, would spend the most creative years of his life.

A series of business failures had weakened Hermann Einstein's earlier robust health. In October, Albert rushed to his father's side in Milan, where Hermann died of a heart condition after a brief illness.

HENRI POINCARÉ, in *La Science et l'hypothèse,* noted that it does not matter whether the ether exists. "What is essential for us is that everything happens as if it existed... It is only a convenient hypothesis... Some day, no doubt, the ether will be thrown aside as useless."

The child at right was thought for a time to be Lieserl.

Bertrand Russell found the "ultimate paradox," the set S of all sets which do not contain themselves. If S *does* contain itself, he maintained, then it cannot belong to the set of all sets which *do not* contain themselves. So S does *not* contain itself. On the other hand, if S does *not* contain itself, then it must belong to the set of all sets that do not contain themselves, so it *does* contain itself. In a letter to mathematician Gottlieb Frege in June, Russell devastated Frege's work on building up numbers from sets, on which Frege had worked since 1879.

The Nobel Prize in physics was awarded jointly to Hendrijk A. Lorentz and Peter Zeeman of the Netherlands for their research into the influence of magnetism on radiation phenomena. In the presence of a magnetic field, the energy levels of an atom are altered. Einstein became a great admirer of Lorentz, both personally and intellectually, calling him "a living work of art." The chemistry prize was awarded to Emil Fischer of Germany for his work on sugar and purine syntheses.

---

2. "On the Thermodynamic Theory of the Potential Difference between Metals and Fully Dissociated Solutions of Their Salts and on an Electrical Method for Investigating Molecular Forces" (Ueber die thermodynamische Theorie der Potentialdifferenz zwischen Metallen und vollständig dissociirten Lösungen ihrer Sälze und über eine elektrische Methode zur Erforschung der Molecularkräfte). *Annalen der Physik* 8 (1902): 798–814.

Einstein discussed the conditions for the validity of the second law of thermodynamics, which became significant for his later work.

---

3. "Kinetic Theory of Thermal Equilibrium and of the Second Law of Thermodynamics" (Kinetische Theorie des Wärmegleichgewichtes und des zweiten Hauptsatzes der Thermodynamik). *Annalen der Physik* 9 (1902): 417–433.

Intending to complete the mechanical foundations of the "general theory of heat," Einstein here provided the keystone in a chain of derivations already begun by Ludwig Boltzmann.

Albert and Mileva Einstein in Bern

1903

Einstein started the new year by getting married. He and Mileva tied the knot in Bern on January 6 in a simple civil ceremony with two wedding guests, Einstein's friends Maurice Solovine and Conrad Habicht, in attendance. The couple began to make their home together in one of the six apartments they would inhabit during their seven years in the Swiss capital.

Bern was a lively and lovely town, with clusters of intellectuals and students gathering in various cafés and homes. The newlyweds fit right in. Einstein's need to express himself freely and mingle socially had already motivated him, Solovine, and Habicht to form what they mockingly called their "Olympia Academy," a discussion group to which they would occasionally invite one or two other friends. This small gathering informally discussed the important scientific and intellectual topics of the day while socializing, sometimes boisterously. The young people debated the philosophical works of Karl Pearson, David Hume, Ernst Mach, Georg Riemann, Baruch

Spinoza, and Henri Poincaré, often well into the night. Mileva was part of the group but remained more a quiet listener than an active participant. Einstein was still trying to find a better job, which was complicated, according to Mileva, by his sharp tongue and being a Jew. She wrote a friend that both of them would even consider teaching German in Budapest if such positions became available.

In September, while Mileva visited her family in Novi Sad, Lieserl came down with scarlet fever. At that time, Einstein wrote to Mileva, asking her how Lieserl was registered, though it is not clear what he meant by "how" *(als was).* Although her birth or baptism may have been registered somewhere, no record of either has ever been found. After this time, neither Albert nor Mileva ever mentioned their daughter again in any surviving correspondence.

In the fall, a year and a half after Lieserl's birth, Mileva wrote Albert from Hungary that she was pregnant again. He wrote back saying he was happy to hear the news and was eagerly awaiting her return home.

Einstein gave his first scientific presentation, "Theory of Electromagnetic Waves," to an association of scientists in Bern in December. And it was perhaps during this year that he began to think about a new topic for a doctoral dissertation.

THE ART AND SCIENCE of transportation was moving forward in America. Most famously, Orville and Wilbur Wright, two bicycle makers from Ohio, launched the first successful manned flight in a motorized airplane. Orville was in the pilot's seat in Kitty Hawk, North Carolina, a small town with the best winds for the occasion. The flight lasted a mere twelve seconds and covered only a hundred feet from start to finish, though the ups and downs in the air made the journey equivalent to 540 feet.

Earlier in the year, a newly built Packard Model F prototype car had already proved it could go much farther. The cross-country road test, on the unpaved roads of a disconnected "highway" system, took fifty-two days from San Francisco to New York City. The same year, Henry Ford founded the Ford Motor Company as the largest car manufacturer in the world and introduced his Model A.

In Europe, Russian physicist K. E. Tsiolkovsky, after building an aeronautical wind tunnel, introduced far-reaching ideas about space travel, including the multi-stage rocket. Dutch physiologist Wilhelm Einthoven invented the string galvanometer, which allowed him to manufacture the electrocardiogram—a graphic record of the beating heart.

Marie and Pierre Curie shared the Nobel Prize in physics with French physicist Henri Becquerel, the latter for his discovery of spontaneous radioactivity and the former for their research on the radiation phenomena discovered by Becquerel. The chemistry prize was awarded to Svante Arrhenius of Sweden for his electrolytic theory of dissociation.

4. "A Theory of the Foundations of Thermodynamics" (Eine Theorie der Grundlagen der Thermodynamik). *Annalen der Physik*, ser. 4, 11 (1903): 170–187.

Einstein showed that the concepts of temperature and entropy follow from the assumption of the energy principle and atomic theory. He required only the foundations of atomic physics but no other physical hypotheses.

The Wright brothers take flight in Kitty Hawk, North Carolina.

1904

On May 14, Albert and Mileva's first son, Hans Albert, was born, and Einstein became a family man at the age of twenty-five. His professional life began to improve when, in September, the provisional appointment as patent clerk at the Patent Office in Bern became permanent and he received a small raise. He became an expert on evaluating electrical gadgets and helped his friend Michele Besso, a mechanical engineer six years his senior, also secure a job at the Patent Office. Besso became Einstein's close and lifelong friend and correspondent, one whom Einstein admired greatly for his personal qualities.

IN THE UNITED STATES, the first railroad tunnel was excavated under the North (now Hudson) River between Manhattan and New Jersey, and the Broadway subway opened, enabling New Yorkers to live farther from work and easing the congestion of the Lower East Side.

In Latin America, work began on the Panama Canal, and W. C. Gorgas, an army physician and sanitation expert, succeeded in controlling yellow fever in the Canal Zone through mosquito-eradication measures.

The Nobel Prize in physics was awarded to Lord Rayleigh (John William Strutt) of the United Kingdom for his investigations of the densities of the most important gases and for his discovery of argon. The Nobel Prize in chemistry was given to Sir William Ramsay, also of the United Kingdom, for discovering the inert gaseous elements in air, such as helium, neon, and krypton, and determining their place in the periodic table.

5. "On the General Molecular Theory of Heat" (Zur allgemeinen molekularen Theorie der Wärme). *Annalen der Physik* 14 (1904): 354–362.

The culmination of Einstein's efforts to generalize and extend the foundations of statistical physics, this was his last paper devoted exclusively to the subject.

Einstein in Bern, Switzerland

1905

The year 1905, Einstein's annus mirabilis, or miracle year in terms of his contributions to physics, established him, at the age of twenty-six, as the world's leading physicist. He not only published five important papers but also found time to write twenty-three review articles for journals. He accomplished all of this on his own time after coming home from the Patent Office, a remarkable achievement for someone not yet in the halls of academia. Despite being outside the academic world, he had no trouble gaining prominence in the world of physics with his innovative and sometimes controversial contributions to the field. Many scholars think that it is precisely because Einstein was unencumbered by the trappings and long hours of academic life that he had the time to think and write clearly and creatively.

Based on a letter from Einstein to Mileva in which he makes a reference to "our work," some historians have suggested that Mileva was Einstein's collaborator in his scientific research. No hard evidence exists for this claim, however. It is known that Albert would run his ideas by her and that she would listen, make suggestions, and probably proofread his papers and point out inconsistencies. But when one evaluates her own background in Einstein's field of concentration, it seems unlikely that she was a creative force behind his work.

In late summer, Einstein took Mileva and Hans Albert on an excursion to Belgrade and Novi Sad to visit Mileva's family and introduce their son to them.

WHILE EINSTEIN was breaking new ground in physics, settling into fatherhood, and getting used to having a steady job in Bern, Sigmund Freud in Vienna startled the world with his *Three Contributions to the Theory of Sex,* a landmark study examining sexual aberrations, infantile sexuality, and the transformations of puberty. Three decades later, Freud and Einstein would have a lively exchange of letters and collaborate on writing a pamphlet, "Why War?" Einstein wrote to a friend that Freud often had an "exaggerated faith in his own ideas." Once, when he was invited to undergo Adlerian psychotherapy, Einstein said that he preferred to remain "in the darkness" and unanalyzed.

In America, the Mount Wilson Observatory was completed near Pasadena in California. Twenty-six years later, in 1931, Einstein would visit the observatory where Nobel laureate Albert A. Michelson had measured stellar diameters and the speed of light. In 1905 Edmund B. Wilson discovered that the X chromosome is linked to the gender of the bearer, and Svante Arrhenius showed his prescience by expressing concern that large-scale burning of fossil fuels might result in global warming.

The Nobel Prize in physics went to Philipp Lenard of Germany for his work on cathode rays. Although Einstein and Lenard had professional respect for each other early in the century, Lenard later became a staunch Nazi and anti-Semite and, in his book *German Physics* (1936), denounced Einstein's "Jewish physics" and relativity the-

Theodore Roosevelt's inaugural parade

ory. "In contrast to the intractable and solicitous desire for truth in the Aryan scientists, the Jew lacks to a striking degree any comprehension of truth," he wrote. Einstein considered Lenard's objections to his theory to be superficial and not worthy of a reply. Also awarded a Nobel Prize, in chemistry, was Adolf von Baeyer, also of Germany, for his work in organic chemistry and on organic dyes and hydroaromatic compounds.

6. "On a Heuristic Point of View Concerning the Production and Transformation of Light" (Über einen die Erzeugung und Verwandlung des Lichtes betreffenden heuristischen Gesichtspunkt). *Annalen der Physik* 17 (1905): 132–148.

Einstein expounded on the peculiar discrepancy between material bodies and radiation and introduced the concept of light quanta, or photons, providing the basis for much of the later work in quantum theory, especially Bohr's theory of the atom. Challenging the wave theory of light, Einstein showed that electromagnetic radiation interacts with matter as if the radiation has a granular structure (the so-called photoelectric effect). He determined that a massless quantum of light, the photon, would have to impart the energy required according to Planck's radiation law to break the attractive forces holding the electrons in the metal. This theory was one of the milestones in the development of quantum mechanics, making Einstein the foremost pioneer in the field and opening the world of quantum physics. The first of the five great papers he published in 1905, it earned him the Nobel Prize in physics sixteen years later.

7. "A New Determination of Molecular Dimensions" (Eine neue Bestimmung der Molekuldimensionen). Dated April 30, 1905, but not published until the following year. Bern: Wyss, 1906. Also slightly revised in *Annalen der Physik* 19 (1906): 289–305.

This document is Einstein's doctoral dissertation, resubmitted in the spring of 1905 after he withdrew his first submission in 1902. Here he combined the techniques of classical thermodynamics with those of the theory of diffusion to create a new method for determining molecular sizes. He wanted to discover facts that would establish once and for all the existence of atoms of a specific finite size, since at the turn of the century the atom's existence was still in contention. After he submitted his copy of the thesis to the University of Zurich, his petition to receive the doctorate was approved.

8. **"On the Motion of Small Particles Suspended in Liquids at Rest Required by the Molecular-Kinetic Theory of Heat" (Über die von der molekularkinetischen Theorie der Wärme geforderte Bewegung von in ruhenden Flüssigkeiten suspendierten Teilchen). *Annalen der Physik* 17 (1905): 549–560.**

For the first time, Einstein discussed Brownian motion, the irregular movement of microscopic particles suspended in a liquid, which was named after the eighteenth-century Scottish botanist who first observed it. As Einstein explained in a letter to his friend Conrad Habicht on May 25, he proved in this paper that particles about 1/1000 millimeter in diameter suspended in liquids and too small to see move randomly because of thermal dynamics. By inverting Boltzmann's formula, Einstein described its mathematics, deriving the probability of a macroscopic state for the distribution of gas molecules. This paper led to experiments validating the kinetic-molecular theory of heat. To date, this is Einstein's most-cited paper.

9. **"On the Electrodynamics of Moving Bodies" (Zur Elektrodynamik bewegter Körper). *Annalen der Physik* 17 (1905): 801–921.**

This landmark in the development of physics, one of the two papers that laid out the theory of special relativity (the other is no. 10), formulated a new conception of time. By assuming that the speed of light is the same to every observer moving at a constant velocity, Einstein showed that space and time were not independent: spacetime was born. According to Hermann Weyl in 1918, this theory "led to the discovery that *time* is associated as a fourth coordinate on an equal footing with the other three coordinates of space, and that the scene of material events, the *world*, is therefore a four-dimensional, metrical continuum." It was a revolutionary piece of scientific work.

Einstein's exploration of the nature of simultaneity and expression of the necessity of *defining* simultaneity led to widespread discussion of the "conventionality of simultaneity" in the philosophy of science, which is still a hot topic today.

Henri Poincaré, also in 1905, obtained, independently from Einstein, many of the results of the special theory.

10. **"Does the Inertia of a Body Depend on Its Energy Content?" (Ist die Trägheit eines Körpers von seinem Energieinhalt abhängig?). *Annalen der Physik* 18 (1905): 639–641.**

Using the postulates of the special theory of relativity, Einstein showed that energy radiated is equivalent to mass lost, which would eventually lead to the famous equation $E = mc^2$. He considered the conservation of energy of a radiating body in a system at rest and in a system in uniform motion relative to it. For the first time he concluded that "the mass of a body is a measure of its energy content." The special theory of relativity paved the way for a deeper appreciation of symmetry criteria in physics and introduced new views on space and time, yet it took twenty-five years for experimental evidence in its favor to emerge. Einstein credited Galileo, Isaac Newton, James Clerk Maxwell, and H. A. Lorentz for laying the foundation for the theory.

1906

In January, Einstein formally received his doctorate from the University of Zurich. Two months later, he was promoted to a higher-rank clerk in the Patent Office in recognition of his ability to analyze difficult patent applications, and his salary also took a jump. By this time, the small but lively Olympia Academy had disbanded because its members had moved to distant places, and Einstein deepened his friendships with other Bernese residents, such as his sister Maja's fiancé, Paul Winteler, and Michele Besso.

After his year of miracles, Einstein eagerly waited for reactions to his publications. Some of the most eminent physicists of the day, such as Nobelist H. A. Lorentz and future Nobelist Max Planck, understood and appreciated the revolutionary implications of his work. Einstein admired Planck greatly, calling him "one of the finest persons I have ever known." Tongue in cheek, Einstein added, "but he really did not understand physics, because during the eclipse of 1919 he stayed up all night to see if it would confirm the bending of light by the gravitational field. If he had *really* understood the general theory of relativity, he would have gone to bed the way I did." He credited Planck's support of his theory for attracting the notice of colleagues in the field so quickly.

Physicists sought Einstein out for scientific discussions, and he expanded his circle of colleagues throughout Europe, pushing Mileva more into the background as physics became increasingly important to his career and happiness. For the next few years, to convince his critics of the validity of his theory, he concentrated on

Earthquake and fire ravage San Francisco.

publishing papers to elaborate his ideas. Still, Einstein was an attentive family man at this time and especially doted on Hans Albert. Mileva and the little boy went to Hungary to spend some time with her parents.

IN AMERICA, the city of San Francisco suffered a destructive earthquake and subsequent fire, resulting in seven hundred deaths and hundreds of millions of dollars in property damage. A more devastating earthquake, not as well known, ravaged Valparaiso, Chile, killing twenty thousand.

Walther Nernst, the German physical chemist and physicist who would win the Nobel Prize in chemistry in 1920, stated a new tenet, often called the third law of thermodynamics: if a chemical change takes place between substances that are at absolute zero, there is no change in entropy.

Roald Amundsen traversed the Northern Passage and established the magnetic North Pole, and Thomas Edison invented the "cameraphone," which would synchronize a phonograph and projector for motion pictures with sound. Frederick Hopkins noticed that "accessory food factors," later called vitamins, were essential to growth in rats. In France, Marie Curie was appointed to her deceased husband and co-Nobelist's position and became the first female professor at the Sorbonne in Paris. Though Einstein admired Curie, behind her back he confessed that he found her "very intelligent but as cold as a herring."

J. J. Thomson of England was awarded the Nobel Prize in physics for his theoretical and experimental investigations on the conduction of electricity by gases, and Henri Moisson of France won the prize for chemistry for his investigation and isolation of the element fluorine and for applying to scientific investigations the electric furnace named after him.

11. **"On the Theory of Brownian Motion" (Zur Theorie der Brownschen Bewegung).** ***Annalen der Physik*** **19 (1906): 371–381.**

Here Einstein presented his earlier ideas on Brownian motion in more elegant form, adding two new applications: the vertical distribution of a suspension under the influence of gravitation and a calculation for a Brownian rotational movement for a rotating solid sphere. Jean Perrin's experiments on the former would win him the Nobel Prize in 1926.

12. **"The Principle of Conservation of Motion of the Center of Gravity and the Inertia of Energy" (Das Prinzip von der Erhaltung der Schwerpunktsbewegung und die Trägheit der Energie).** ***Annalen der Physik*** **20 (1906): 627–633.**

In an ingenious thought experiment involving energy transport in a hollow cylinder, Einstein returned to the relationship between inertial mass and energy, giving more general arguments for their complete equivalence.

1907

Einstein's work was now being discussed seriously by the most prominent physicists in Europe. In June, however, when he applied for a postdoctoral position at the University of Bern, his application was turned down because he did not submit the requisite unpublished thesis *(Habilitationsschrift)*. In its place he had sent the search committee a bundle of his already published papers, feeling that these should suffice.

In the fall, after contemplating why relativity seemed to apply to almost every physical phenomenon except gravity, he formulated the principle of equivalence for uniformly accelerated mechanical systems. While sitting in the Patent Office, he suddenly realized that, if a person were in a free fall, he would not feel his own weight. Because everything falling with the person falls at the same rate, there is no way to tell that he is in a gravitational field—there is no reference point. Einstein concluded that the person could assume that he was at rest and everything around him was being pulled upward—that is, gravity seemed to be relative. This idea, after he thought about it more deeply, would lead Einstein on the path to the theory of general relativity. He called it "the happiest thought of my life." At this time, Einstein also became interested in the unexplained motions of the planet Mercury.

THE AMERICAN ENGINEER George W. Goethals was appointed to direct the construction of the Panama Canal.

The Nobel Prize in physics was awarded to the American Albert Michelson for his optical precision instruments and the spectroscopic and metrological investigations carried out with their aid. Einstein would later say that he thought of Michelson as "the artist in science. His greatest joy seemed to come from the beauty of the experiment itself and the elegance of the method employed." The chemistry prize went to Eduard Buchner of Germany for his biochemical research and his discovery of cell-free fermentation.

13. "Planck's Theory of Radiation and the Theory of Specific Heat" (Die Plancksche Theorie der Strahlung und die Theorie der spezifischen Wärme). *Annalen der Physik* 22 (1907): 180–190.

In the first paper he wrote on the quantum theory of solids, Einstein made a deduction of Planck's radiation formula and systematically introduced probability factors in the mathematics of quantum theory. About three months later, he published a short erratum on the paper. A classic paper, it offered a complete explanation for the specific heat of solids from absolute zero to above room temperature.

14. **"On the Limit of Validity of the Law of Thermodynamic Equilibrium and on the Possibility of a New Determination of the Elementary Quanta" (Über die Gültigkeit des Satzes vom thermodynamischen Gleichgewicht und über die Möglichkeit einer neuen Bestimming der Elementarquanta).** ***Annalen der Physik*** **22 (1907): 569–572.**

Einstein used the thermodynamic approach to fluctuations in Brownian motion to predict voltage fluctuations in condensers. To test his theory, he needed a new, highly sensitive instrument—more sensitive than the available electrometers, which could measure to a few thousandths of a volt. Einstein designed it, had it built, and famously called it his *Maschinchen* (little machine). He toyed with the idea of patenting it but then dismissed the notion when manufacturers showed little interest. Instead, he decided to publish a paper on the basic features of his machine the following year.

15. **"Theoretical Remarks on Brownian Motion" (Theoretische Bemerkungen über die Brownsche Bewegung).** ***Zeitschrift für Elektrochemie und angewandte physikalische Chemie*** **13 (1907): 41–42.**

Einstein attempted to make the fundamental features of his theory accessible to readers who had only a moderate background in mathematics. He discussed some peculiarities of the statistical motion of particles suspended in a fluid that can hamper experimental verification.

16. **"On the Inertia of Energy Required by the Relativity Principle" (Über die vom Relativitätsprinzip geforderte Trägheit der Energie).** ***Annalen der Physik*** **23 (1907): 371–384.**

As in paper 12 above, Einstein discussed the relationship between inertial mass and energy, arguing for their complete equivalence, namely, that every mass has an equivalent energy just as every form of energy has an equivalent mass. This relation says that a photon can convert into matter with the appropriate mass, and vice versa. He deduced the exact expression for the equivalence of mass and energy, his celebrated equation $E = mc^2$. He also returned to the question of the impossibility of superluminal velocities.

17. **"On the Relativity Principle and the Conclusions Drawn from It" (Über das Relativitätsprinzip und die aus demselben gezogenen Folgerungen).** ***Jahrbuch der Radioaktivität und Elektronik*** **4 (1907): 411–462. Einstein published some corrections to the paper the following year in vol. 5 (1908): 98–99.**

In this review article on relativity, Einstein summarized the results of some of his earlier papers on the theory of relativity, sometimes simplifying his earlier proofs. It covered relativistic kinematics, optics, electromagnetic theory, and the relativistic dynamics of a particle and of an extended system. He proved that a body's inertial and gravitational masses are equal to the same quantity E/c^2 and therefore should be considered precisely equal to each other. This result would be a steppingstone to the 1915 theory on general relativity.

1908

Einstein finally submitted the required thesis (which still remains unpublished because it was apparently discarded) that would qualify him for a lecturer's position at the University of Bern. He was hired the month after its submission, delivered his inaugural lecture at the university on February 27, and commenced his teaching duties in late April, with a course on the molecular theory of heat. After the course ended in late July, he headed to the Bernese mountains for a vacation with Mileva and Hans Albert.

Throughout the year, Einstein, his friends the Habicht brothers, and to some extent Mileva were intensely involved with continuing experiments on his *Maschinchen,* for which they eventually received a patent. He continued to enjoy tinkering with gadgets and solving puzzles all his life. When the winter semester began in October, he was assigned to teach a course on the theory of radiation. At the end of the year, his sister Maja was awarded a doctorate in romance languages from the University of Bern.

HERMANN MINKOWSKI took Einstein's algebraic expression of the special theory of relativity and geometrized it, combining space and time into a four-dimensional continuum; he thereby provided a framework for all future work in relativity. In California, George Ellery Hale and his team completed installing the sixty-inch reflecting telescope at Mount Wilson. Robert Millikan, whom Einstein would meet later at the California Institute of Technology in Pasadena, determined the probable minimum unit of an electrical charge, or the electron (or "beta particle," as the electron was called at that time). He later coined the term *cosmic rays.*

Two of Einstein's future friends, German chemist Fritz Haber and Dutch physicist Heike Kamerlingh Onnes, synthesized ammonia and liquefied helium, respectively. Haber's synthesis of ammonia eventually led to the Oppau and Leuna Ammonia Works, which enabled Germany to prolong World War I after its supply of nitrogen for making explosives ran out in 1914. Both Kamerlingh Onnes and Haber became Nobel laureates, in 1913 and 1918, respectively. Meanwhile, English electrical engineer A. A. Campbell Swinton proposed the first all-electronic scheme for television using a cathode ray tube. His system was never built, and interest in television languished for several years.

Early on the morning of June 30, a mysterious explosion occurred over the Siberian sky: a large meteorite broke up at an altitude of about six kilometers, causing a brilliant fireball and deafening bang. Because the meteor exploded in the atmosphere, it left no crater, but the force of the blast devastated a large area of Siberian forest and killed hundreds of reindeer and one herder.

Gabriel Lippmann of France won the Nobel Prize in physics for his method of reproducing colors photographically based on the phenomenon of interference. Ernest Rutherford, a New Zealander living in England, was awarded the prize in chemistry for his investigations into the disintegration of the elements and the chemistry of radioactive substances.

18. "A New Electrostatic Method for the Measurement of Small Quantities of Electricity" (Eine neue elektrostatische Methode zur Messung kleiner Elektrizitätsmengen). *Physikalische Zeitschrift* 9 (1908): 216–217.

Einstein published the basic features of his method for constructing the *Maschinchen* that he used in his work on paper 14.

19. "Elementary Theory of Brownian Motion" (Elementare Theorie der Brownschen Bewegung). *Zeitschrift für Elektrochemie und ungewandte physikalische Chemie* 14 (1908): 235–239.

An elementary account of Einstein's theory of Brownian motion was needed, especially by chemists who were trying to provide experimental support for it and often misunderstood it. Einstein elaborated on the relation between diffusion and osmotic pressure and calculated the diffusion coefficient from the frictional resistance of the solvent fluid to the dissolved molecules.

20. "On the Fundamental Electromagnetic Equations for Moving Bodies" (Über die elektromagnetischen Grundgleichungen für bewegte Körper) (with Jakob Laub). *Annalen der Physik* 26 (1908): 532–540. A correction to this paper was published later in the year in vol. 27 (1908): 232, and a supplement the following year in vol. 28 (1909): 445–447.

Elaborating on the relativistic transformation of Maxwell's equations for vacuum discussed in paper 8 above, Einstein also considered the displacement vector **D** and the magnetic induction **B.**

21. "On the Ponderomotive Forces Exerted on Bodies at Rest in the Electromagnetic Field" (Über die im elektromagnetischen Felde auf ruhende Körper ausgeübten ponderomotorischen Kräfte) (with Jakob Laub). *Annalen der Physik* 26 (1908): 541–550.

Einstein wrote papers 20 and 21 in a three-week period with Jakob Laub, his first scientific collaborator, to address the problems of formulating relativistically invariant equations for electromagnetic fields in moving media, first raised by Hermann Minkowski the year before. Laub was also involved in making the corrections and writing the supplement.

1909

Einstein's reputation as a young physicist in his prime, full of innovative ideas, continued to spread, giving him the credentials to meet and become better acquainted with leading European physicists. Consequently, his ambitions to become an academic himself were finally realized. Upon his appointment as associate professor of physics at the University of Zurich, to begin in October, he tendered his resignations to the Patent Office and the University of Bern. His starting salary at Zurich, however, was the same as the salary he had been earning as a patent clerk, a situation he hoped would soon change.

That summer, two days after submitting his resignation, he received his first honorary doctorate, from the University of Geneva (in his lifetime he would receive approximately twenty-five such honors). Characteristically, he made light of the occasion, being the only man to arrive in a summer suit and straw hat, while other dignitaries and recipients decked themselves out lavishly in their finest clothes or academic robes. In October, he moved his family to Zurich. Finding an apartment was difficult due to a housing shortage caused by a labor strike that lasted several years.

Mileva was hesitant to leave Bern and became increasingly insecure about Albert's fame and its effect on the family. He was often called away from her and Hans Albert. She wrote to a friend, "I only hope and wish that fame does not have a harmful effect on his humanity." She became worried about the attention he received from other women, feeling jealous, angry, and neglected. At the end of the summer, before Albert gave a much-anticipated lecture at a physics conference in Salzburg, she had been able to convince him to take a vacation together in the southern Alps. The excursion seemed to mend some romantic fences: by the end of October, shortly after Einstein assumed his teaching duties at the University of Zurich, Mileva was pregnant again. As her pregnancy advanced, Einstein became engrossed in teaching a course in mechanics and another course in thermodynamics and presided over a physics seminar.

WOMEN AT LAST began to be admitted to German universities; previously, only selected foreign women had been allowed to study there. It was still difficult for women to get postgraduate degrees, however.

Sigmund Freud came to the United States to lecture on psychoanalysis. American Robert E. Peary reached the North Pole, and the first commercial manufacturing of Bakelite marked the beginning of the Plastic Age. Incandescent lights were used for the first time in automobile headlights, replacing carbide flame jets.

German chemist Fritz Haber developed the first commercially important high-pressure chemical process to synthesize ammonia from its elements. Meanwhile, Hans Geiger, of Geiger counter fame, and Ernest Marsden, under the direction of Ernest Rutherford, published their paper describing how they scattered alpha particles with thin films of heavy metals, providing evidence that atoms have a discrete nucleus. Rutherford, who had won the Nobel Prize in chemistry in 1908, began a systematic theoretical investigation of Geiger and Marsden's results, as well as those of similar experiments with beta particles (electrons). These studies culminated in a classic paper (1911) showing that the observations did not agree with the then-current picture of the atom as a soft, jellylike sphere with diffusely distributed positive and negative charges (the "plum pudding" model).

German mathematician David Hilbert's work on integral equations established the basis for his subsequent work on infinite-dimensional space, or "Hilbert space." Mathematical investigations of quantum mechanics usually make use of the properties of Hilbert space. Hilbert had, in 1900 at the Paris International Congress of Mathematicians, proposed twenty-three mathematical problems—which came to be known as "Hilbert problems"—that he thought should be solved during the twentieth century. Some of these problems remain totally or partially unsolved today.

The Nobel Prize in chemistry went to Wilhelm Ostwald of Germany in recognition of his work on catalysis and his investigations into the fundamental principles governing chemical equilibria and rates of reaction. The prize in physics was awarded jointly to Guglielmo Marconi and Carl Ferdinand Braun for their contributions to the development of wireless telegraphy. Braun had modified Marconi's transmitters, significantly increasing their range and usefulness.

22. **"On the Present Status of the Radiation Problem" (Zum gegenwärtigen Stand des Strahlungsproblem). *Physikalische Zeitschrift* 10 (1909): 185–193. In another article of the same title published in the same volume, pp. 323–324, Einstein and Walter Ritz summarize their differences on the radiation problem.**

Responding to papers published by H. A. Lorentz, James Jeans, and Walter Ritz the preceding year, in which they discussed their respective opinions on the radiation problem, Einstein elaborated the concept of the "statistical probability of a state" first introduced in paper 6 above. The novelties in this paper are the two arguments for the existence of light quanta based on the analysis of fluctuations in black-body radiation. After making his first-known attempt to find a field theory that would explain the structure of both matter and radiation, Einstein admitted that he had not yet succeeded in finding a system of equations for both. This attempt was a forerunner of his later search for a unified field theory.

23. "On the Development of Our Views Concerning the Nature and Constitution of Radiation" (Über die Entwicklung unserer Anschauungen über das Wesen und die Konstitution der Strahlung). *Deutsche Physikalische Gesellschaft. Verhandlungen* 7 (1909): 482–500. Also published the following month in *Physikalische Zeitschrift* 10 (1909): 817–825.

Einstein had presented this lecture in September to a meeting of the physics section of an association of German scientists and physicians held in Salzburg, Austria, the first such conference he attended. He masterfully summarized his views on radiation and for the first time publicly linked his work on relativity and the quantum hypothesis. This is the first synthesis showing the profound changes in the concept of light brought about by the theory of relativity and the important implications this change would have on the development of physics. He reiterated that light has an independent existence, just like matter.

1910

Einstein's fame continued to grow throughout Europe as he published his ideas and review articles, taught his courses, and delivered lectures to scientific audiences. Not surprisingly, many institutions became interested in having him on their faculties. In the spring, he was proposed for a chair in theoretical physics at the German University of Prague. His consideration of the offer prompted his enthusiastic and loyal students to petition the officials in Zurich to encourage him to stay and offer him a raise. Perhaps unconvinced that the offered increase of one thousand Swiss francs per year was enough, Einstein traveled to Vienna in the early fall to discuss the Prague appointment with the Habsburg authorities. Even though Einstein was not a brilliant teacher, his students enjoyed his down-to-earth congeniality. In demand were not his lecturing skills but his creative ideas and his ability to engage his students in informal scientific socializing.

The year had already been full of family milestones. Einstein's sister, Maja, had married Paul Winteler in March, Mileva's parents had come to visit in the summer, and in July his second son, Eduard ("Tete," or Teddy), was born. Such events were now taking second place in Einstein's life behind his scientific and professional pursuits. Mileva complained, "With that kind of fame he does not have much time left for his wife."

Halley's comet whips by the earth.

ERNEST RUTHERFORD'S investigations into the scattering of alpha particles and the nature of the inner structure of the atom that caused such scattering led to his theory of a "nucleus," his greatest contribution to physics. He postulated that almost the whole mass of the atom and its positive charges are concentrated in a minute space at the center. He published his findings the following year.

The French engineer, chemist, and inventor Georges Claude displayed the first neon lamp to the public in Paris. Claude had discovered that, when there is a voltage drop across a tube of neon gas, a bright red glow emerges and that other gases give off different colors; for example, argon gives off blue, and helium gives off yellow and white. Thirteen years later, he and his French company introduced the first neon signs to the United States, bought by a Packard auto dealership in Los Angeles.

Thomas Hunt Morgan proposed the gene theory of heredity. He would receive the Nobel Prize in physiology and medicine in 1933 for his discoveries of the role played by the chromosome in heredity.

The Nobel Prize in physics was awarded to Johannes van der Waals of the Netherlands for his work on the equation of state for gases and liquids. The chemistry prize went to Otto Wallach of Germany in recognition of his services to organic chemistry and the chemical industry through his pioneering work in the field of alicyclic compounds.

24. **"The Principle of Relativity and Its Consequences in Modern Physics" (Le Principe de relativité et ses conséquences dans la physique moderne).** ***Archives des sciences physiques et naturelles*** **29 (1910): 5–28 and 125–144.**

Published as two separate papers in the same journal and translated from the German by Edouard Guillaume, this work is a general survey of the history and essence of the theory of relativity and its applications. In a letter to Jakob Laub of August 27, Einstein stated that the work did not contain any new insights, that it "merely comprises a rather general discussion of the epistemological foundations of the theory of relativity, no new views whatsoever, and almost nothing that is quantitative."

25. **"On the Theory of Light Quanta and the Question of the Localization of Electromagnetic Energy" (Sur la théorie des quantités lumineuses et la question de la localisation de l'énergie électromagnétique).** ***Archives des sciences physiques et naturelles*** **29 (1910): 525–528.**

Einstein presented this paper in May at a meeting of the Swiss Physical Society in Neuchâtel. Much of it is based on his earlier work on the quantum hypothesis, which cannot be reconciled with the accepted theory of radiation.

26. **"On the Ponderomotive Forces Acting on Ferromagnetic Conductors Carrying a Current in a Magnetic Field" (Sur les forces pondéromotrices qui agissent sur des conducteurs ferromagnétiques disposés dans un champ magnétique et parcourus par un courant).** ***Archives des sciences physiques et naturelles*** **30 (1910): 323–324.**

Einstein presented this paper to a meeting of the Schweizerische Naturforschende Gesellschaft in Basel on September 6, 1910. He discussed the question of a ponderomotive force exerted on a ferromagnetic substance in the presence of an external magnetic field, coming up with the only expression that satisfies the principle that action equals reaction.

27. **"On a Theorem of the Probability Calculus and Its Application in the Theory of Radiation" (Über einen Satz der Wahrscheinlichkeitsrechnung und seine Anwendung in der Strahlungstheorie) (with Ludwig Hopf).** ***Annalen der Physik*** **33 (1910): 1096–1104.**

Einstein wrote this paper to show that the failure of statistical mechanics vis-à-vis the radiation law cannot be ameliorated by proposing that individual statistical events should not follow the usual law of independence instead of assuming a certain interdependence between them.

28. "Statistical Investigation of a Resonator's Motion in a Radiation Field" (Statistische Untersuchung der Bewegung eines Resonators in einem Strahlungsfeld) (with Ludwig Hopf). *Annalen der Physik* 33 (1910): 1105–1115.

The authors make use of the results in paper 27, demonstrating that the Rayleigh-Jeans law of radiation is an unavoidable consequence of statistics, even if we avoid assumptions that we may think need correction. In other words, we cannot blame statistics for a faulty result.

29. "The Theory of Opalescence of Homogeneous Fluids and Liquid Mixtures near the Critical State" (Theorie der Opaleszenz von homogenen Flüssigkeiten und Flüssigkeitsgemischen in der Nähe des kritischen Zustandes). *Annalen der Physik* 33 (1910): 1275–1298.

Einstein explained the optical effects that occur near the critical point of a fluid (at which liquid and gas phases can coexist) and of a binary mixture of liquids, which can also explain the blue color of the sky. Adding to earlier studies that provided evidence for the atomistic constitution of matter, this is one of his most difficult papers to understand.

1911

In early January, Einstein received the news that Emperor Franz Joseph of the Austro-Hungarian (Habsburg) Empire had officially appointed him to the chair of theoretical physics at the German University in Prague, effective in April. Einstein would also be the director of the Institute of Theoretical Physics, a highly prestigious position. He would teach courses in mechanics and thermodynamics that year and lead a physics seminar. The salary was about twice the amount the Swiss could afford to pay, making this a position Einstein could not turn down. Later that month, he sent his letter of resignation to the Zurich officials. At the end of March, he moved his family to the charming capital city.

With the new salary, Albert and Mileva were able to afford a more bourgeois lifestyle in a spacious apartment, with enough room for a maid. Still, Mileva was not happy with the move, as Prague was known to be home to a snobby and pompous society that looked down their noses on all but the Germans. Mileva started to show signs of serious depression, a condition that had plagued other members of her family. Later, when Eduard entered adulthood, he was diagnosed with schizophrenia

SS *Titanic* nears completion. More than fifteen hundred people will die on its maiden voyage.

(until then known as *dementia praecox*). Despite the Einsteins' more affluent lifestyle, the mood in the household became increasingly gloomy.

During 1911, Einstein received job offers from the University of Utrecht and the Swiss Federal Polytechnical School in Zurich, and he negotiated to secure a new position the following year. Among those recommending him for the Poly position were Marie Curie and the well-known French mathematician and philosopher of science Henri Poincaré, the creator of topology and founder of the mathematical theory of dynamic systems.

At this time, Einstein first experienced the serious stomach problems that would plague him throughout his life. Still, he became a regular visitor to one of the salons of the old city, run by Berta and Otto Fanta; forty years later, their niece Johanna would become an intimate friend of Einstein's in Princeton. At the Fantas', young Jewish intellectuals gathered weekly to chat about philosophy and science, much as Einstein had done in earlier years with his pals in Bern. In November he delivered a lecture at the first Solvay Congress, a congregation of the world's leading physicists, in Brussels (see paper 34 below). Einstein was the youngest participant and became acquainted with Marie Curie.

Einstein formulated his first decisive ideas on general

relativity—the effect of gravity on light—and suggested that his theory be tested during a total solar eclipse.

MARIE CURIE won her second Nobel Prize, this one for advancing chemistry by discovering the elements radium and polonium. The prize in physics was awarded to Germany's Wilhelm Wien, recognizing his work on the laws governing radiation, valid for short wavelengths.

Ernest Rutherford, thinking about the nature of nuclei that could produce radiation, published the paper, "The Scattering of Alpha and Beta Particles by Matter and the Structure of the Atom," in which he described the atom as a small, heavy nucleus surrounded by electrons, putting on paper his investigations of 1910. Working under Rutherford, Germany's Hans Geiger introduced the first Geiger counter, capable of detecting and counting alpha rays. And Heike Kamerlingh Onnes of the Netherlands discovered superconductivity, or the ability of certain materials to carry electric currents without resistance at low temperatures.

Captain Roald Amundsen and his crew are the first to reach the South Pole.

30. "A Relationship between Elastic Behavior and Specific Heat in Solids with a Monatomic Molecule" (Eine Beziehung zwischen dem elastischen Verhalten und der spezifischen Wärme bei festen Körpern mit einatomigem Molekul). *Annalen der Physik* 34 (1911): 170–174.

To supplement William Sutherland's observation that infrared eigenfrequencies of solid bodies possibly originate in the elastic vibrations of these bodies, Einstein added that electrically charged ions are the source of optical vibrations, while elastic vibrations are caused by the mutual motions of the entire molecule.

31. "The Theory of Relativity" (Die Relativitätstheorie). *Naturforschende Gesellschaft in Zürich. Vierteljahrsschrift* 56 (1911): 1–14.

Einstein presented this paper at a January meeting of the Naturforschende Gesellschaft (Society of Natural Scientists) in Zurich as a farewell lecture after resigning from the University of Zurich to go to Prague. For the first time, he used the term *relativity theory* in a paper, having earlier felt that *relativity principle* was a more accurate description. In the exposition of the concepts and principles of special relativity, this paper is similar to paper 24 above, though it is less technical.

32. "Elementary Observations on Thermal Molecular Motion in Solids" (Elementare Betrachtungen über die thermische Molekularbewegung in festen Körpern). *Annalen der Physik* 35 (1911): 679–694.

Here Einstein continued the work he had begun in 1907 on the specific heat of solids. The heat agitation of solids was reduced to a monochromatic oscillation of the atom, and the specific heat was determined based on the quantum treatment of an oscillator in a radiation field. He explained the discrepancies between his formula and measurements at low temperatures.

33. "On the Influence of Gravitation on the Propagation of Light" (Über den Einfluss der Schwerkraft auf die Ausbreitung des Lichtes). *Annalen der Physik* 35 (1911): 898–908.

Einstein returned to his thoughts on gravitation and discussed his ideas on the static gravitational field, advancing the "half-shift" prediction. In early papers on the subject (see also papers 36 to 38 below), he used two important features: the principle of equivalence and the role of the speed of light. In this paper Einstein took a broader perspective, saying that if a light beam is bent in an accelerating frame of reference, then (if the theory is correct) it must also be bent by gravity by the equivalent amount. His prediction of the bending of light by a gravitational field was one of the key tests of general relativity.

34. **"On the Present State of the Problem of Specific Heats" (Zum gegenwärtigen Stande des Problems der spezifischen Wärme). Paper presented at the first Solvay Congress, November 3, 1911. Published later in Arnold Eucken, ed., *Die Theorie der Strahlung und der Quanten* (Halle a.s.: Knapp, 1914), 330–352.**

In this report to an international congress, Einstein elaborated in detail his multifaceted ideas involving the theory of quanta.

1912

Einstein began to prepare a manuscript—as a contribution to a handbook on radiology—reviewing what later came to be called the "special theory" of relativity (to distinguish it from the 1915–16 "general theory"). It took him two years to finish the work, but then World War I intervened and interrupted its publication. The editor asked him to revise and update the work several years later, but at that point he was too busy. It was finally published in volume 4 of *The Collected Papers of Albert Einstein* and gives valuable insights into Einstein's ideas on relativity until that time.

In late January, Einstein's appointment as professor of theoretical physics at the Poly (since 1911 known as

Revolution sweeps Mexico. Peasants rebel under Pancho Villa and Emiliano Zapata.

the Federal Institute of Technology, or Eidgenössische Technische Hochschule—ETH) became official, and in July he and the family happily left Prague to move back to Zurich. Before he moved, Einstein traveled to Berlin, where members of the extended Einstein family were living. He paid his respects to his aunt and uncle and became reacquainted with his cousin Elsa, whom he remembered from childhood visits. Elsa, three years older than Albert, was now divorced and had two daughters: Ilse, who was thirteen, and eleven-year-old Margot. Elsa's and Albert's mothers were sisters, and their fathers were first cousins. Elsa, with her light blue eyes and sunny and outgoing disposition, was the total opposite of Mileva, and Einstein was immediately attracted to the vital and energetic woman. He returned to Prague and began a secret correspondence with her, which continued after he and Mileva moved back to Zurich. Mileva, meanwhile, wrote to friends that she was happy to be back in Zurich and away from the unhygienic and sooty conditions of Prague, which had had adverse effects on her children's health. Regarding her husband, she confessed "with a bit of shame that we are unimportant to him and take second place."

Though Einstein enjoyed returning to the tight circle of his Zurich friends, Mileva became more downcast and morose and began having difficulty walking because of pain in her legs. Einstein distanced himself even more from her, often escaping to play the violin in small ensembles of friends.

The Balkan states, including Serbia, Mileva's homeland, began the First Balkan War, causing Mileva concern for her friends and family in Belgrade.

EINSTEIN'S FRIEND Max von Laue obtained the first diffraction effects by letting x-rays fall on a crystal, causing William Bragg to propose a simple relationship between an x-ray diffraction pattern and the arrangement of atoms in a crystal that produced the pattern, thus inventing x-ray crystallography. Two years later, von Laue would receive the Nobel Prize in physics for 1914.

Carl Jung conceptualized and used the terms *introvert* and *extrovert* when he published his *Theory of Psychoanalysis.*

On the pseudoscientific front, great excitement was engendered when the remains of Piltdown man, believed to be fifty thousand years old, were found in England by amateur geologist Charles Dawson. Many years later, it was established that the bones were those of a modern human and an orangutan, faked to suggest great age, planted by someone as a hoax. Several of the most eminent British scientists of the day had proclaimed the specimens as a genuine "missing link."

The Nobel Prize in physics was awarded to Nils Gustaf Dalen of Sweden for his invention of automatic regulators to be used with gas accumulators for illuminating lighthouses and buoys. The prize in chemistry was divided between two Frenchmen: Victor Grignard, for discovering the "Grignard reagent," and Paul Sabatier, for his method of hydrogenating organic compounds in the presence of finely disintegrated metals, both advancing the progress of organic chemistry.

35. "Thermodynamic Proof of the Law of Photochemical Equivalence" (Thermodynamische Begründung des photochemischen Äquivalentgesetzes). *Annalen der Physik* 37 (1912): 832–838.

Einstein presented a continuation of his earlier work on the interaction between light and matter and on photochemical processes but made no use of the quantum hypothesis. He demonstrated how "the law of photochemical equivalence" is deducible by purely thermodynamic arguments if one makes certain plausible assumptions. He wrote a supplement to the paper five months later in the same journal.

36. "The Speed of Light and the Statics of the Gravitational Field" (Lichtgeschwindigkeit und Statik des Gravitationsfeldes). *Annalen der Physik* 38 (1912): 355–369.

Further exploring his studies of gravitation, based on the equivalence principle, Einstein saw with growing clarity that gravitation is intimately linked with the problem of space and time.

37. "On the Theory of the Static Gravitational Field" (Zur Theorie des statischen Gravitationsfeldes). *Annalen der Physik* 38 (1912): 443–458.

Einstein more closely analyzed the equations of motion stated in paper 36, concluding that those equations cannot be reconciled with the given field equations for c because the principle of "action equals reaction" is violated. He modified the field equation for c. This paper includes the first rudimentary statement of the "law of the geodesics," which was important in the final formulation of the gravitation theory.

38. "Is There a Gravitational Field That Is Analogous to Electrodynamic Induction?" (Gibt es eine Gravitationswirkung, die der elektrodynamischen Induktionswirkung Analog Ist?). *Vierteljahrsschrift für gerichtliche Medizin* 44 (1912): 37–40.

This paper shows the first steps that Einstein took to go beyond his static theory as he proceeded systematically from the special to the more general theory of relativity. It may be surprising that he wrote this paper for a forensic medical journal, but this was no doubt because of the influence of his friend, Heinrich Zangger, the department head and professor of forensic medicine at the University of Zurich. Einstein did attend conferences at which doctors were present, such as the Society of German Scientists and Physicians in September 1911 in Karlsruhe, where he had taken part in discussions following several lectures.

1913

Demanding the right to vote, suffragettes march on Washington, D.C.

The Einstein marriage continued to deteriorate as Mileva became incapacitated by dreadful pain in her legs. Not to be distracted, Albert kept busy with work on his new gravitational theory and tried to avoid thinking about Elsa. He had stopped corresponding with her for nine months but resumed writing after she sent him a birthday letter. Later in the spring, he and Mileva managed to make time for a trip to Paris, where Albert gave a lecture and the couple were houseguests of the widowed Marie Curie, now the holder of two Nobel prizes.

In late summer, Albert and Mileva went to the Engadine Valley of the Swiss Alps with Curie and her daughters. Mileva was unable to hike with the others but enjoyed the fresh air and scenery. In September, the Einsteins took the boys to see their grandparents in Hungary, where Hans Albert and Eduard were baptized in the local Serbian Orthodox church at Mileva's request. Their father, who had no interest in organized religion of any kind, did not attend the baptism. The Einsteins went on to Vienna, where Albert's talk on his new gravitational theory thrust it into the mainstream newspapers. Mileva returned home to Zurich while Albert made a trip to Berlin, where, besides attending to some business, he also met secretly with Elsa.

Einstein published a paper on general relativity theory with his friend, mathematician Marcel Grossmann, as coauthor. Einstein explained the physical principles, and Grossmann wrote the section on advanced mathematics.

In November, Einstein was elected to the Prussian Academy of Sciences and was offered a research professorship at the University of Berlin and the directorate of the soon-to-be-established Kaiser Wilhelm Institute of Physics. He accepted the offer.

While the family was making plans to move to Berlin, the Second Balkan War was spreading to southeastern Europe. Serbia, Mileva's home, was among these war-torn countries.

IN THE WORLD OF SCIENCE, Niels Bohr formulated his theory of atomic structure in his classic paper, "On the Constitution of Atoms and Molecules." This was the first exposition of the theory of quantum mechanics, and in it Bohr viewed atoms as similar to the Sun and planets, with the nucleus at the center and electrons orbiting around it. Bohr explained that the atom absorbs and emits energy by leaping from one orbit to another without traversing the space in between. Meanwhile, Oxford physicist Henry Moseley investigated the x-ray spectra of the elements. Sadly, he was killed in action in 1915 at Gallipoli.

French physicists Charles Fabry and Henri Buisson reported the existence of ozone, a gas created by a photochemical reaction between sunlight and oxygen. Today we know that the layer of ozone around the earth has, over millions of years, played a significant role in the evolution of life on Earth by protecting our planet from the damaging ultraviolet rays of the Sun.

Albert Schweitzer, a future Einstein correspondent

Work on the Panama Canal begins.

and someone whom Einstein admired, opened his hospital in the French Congo at Lambaréné to bring modern medical care into the African jungle.

The Nobel Prize in physics was awarded to Heike Kamerlingh Onnes of the Netherlands for his investigations of the properties of matter at low temperatures, which led, among other things, to the production of liquid helium. Kamerlingh Onnes, who was on friendly terms with Einstein, was professor of experimental physics at the University of Leiden and director of its Institute of Experimental Physics. The chemistry prize went to Alfred Werner of Switzerland for his work on bonding relations of atoms in molecules, which opened up new fields of research, especially in organic chemistry.

39. "Thermodynamic Deduction of the Law of Photochemical Equivalence" (Deduction thermodynamique de la loi de l'equivalence photochimique). *Journal de physique* 3 (1913): 277–282.

This is a published version of a lecture Einstein gave to the Societé Française de Physique at the end of March, dealing with the topics covered in the supplement to paper 35 above.

40. *Outline of a Generalized Theory of Relativity and of a Theory of Gravitation* (Entwurf einer verallgemeinerten Relativitätstheorie und einer Theorie der Gravitation) (with Marcel Grossmann). Leipzig: Teubner, 1913.

In this book, Einstein and Grossmann investigated curved space and curved time as they relate to a theory of gravity. They presented virtually all the elements of the general theory of relativity with the exception of one striking omission: gravitational field equations that were not generally covariant. Einstein soon reconciled himself to this lack of general covariance through the "hole argument," which sought to establish that generally covariant gravitational field equations would be physically uninteresting. Einstein did not adopt the gravitational field equations until late in 1915 in his final formulations of the general theory. Here, Einstein contributed the physics and Grossmann the mathematics. When Einstein moved to Berlin in 1914, his and Grossmann's collaboration ended. Also see paper 44 below on the hole argument.

41. "Physical Foundations of a Theory of Gravitation" (Physikalische Grundlagen einer Gravitationstheorie). Lecture given on September 9, 1913, to the Ninety-sixth Annual Meeting of the Schweizerische Naturgesellschaft (Swiss Society of Natural Scientists) in Frauenfeld, not published until the following year in *Naturforschende Gesellschaft in Zürich. Vierteljahrsschrift* 58 (1913): 284–290.

In this talk, Einstein gave a concise and clear presentation of his gravitational theory. He emphasized the physical content and omitted the advanced mathematical details, discussing the evolution from special to general covariance and showing that the gravitational equations remain outside the framework of general covariance.

42. "On the Present State of the Problem of Gravitation" (Zum gegenwärtigen Zustande des Gravitationsproblems). *Physikalische Zeitschrift* 14 (1913): 1249–1262.

This is the published version of the lecture Einstein gave at the Eighty-fifth Meeting of the Gesellschaft Deutscher Naturforscher und Ärzte (Congress of German Natural Scientists and Physicians) in Vienna on September 23, 1913. In addition to discussing his own work, particularly paper 40 above, he covered Gunnar Nordström's scalar theory, a serious competitor to Einstein's theory. He concluded that only experience could show which of the two was correct. Further discussion by participants followed the talk.

43. "Max Planck als Forscher" (Max Planck as Scientist). *Die Naturwissenschaften* 1 (1913): 1077–1079. See also *The Collected Papers of Albert Einstein*, vol. 4, *Writings, 1912–1914*, 561–563.

Einstein wrote of German physicist Planck as an artist as well as a scientist, explaining that his scientific achievements show artistic creativity. He would write about Planck again in 1918 on Planck's sixtieth birthday and on the occasion of his death in 1948.

1914

At the dawn of World War I in Europe, Einstein began his "Berlin years." People now began to approach him for his opinions on nonscientific subjects as well, mostly because he was so outspoken about his pacifism.

Many physicists continued to be skeptical about relativity theory because they did not have the necessary tools and equations to understand it and no one had proven it experimentally. Einstein continued to clarify his ideas in his publications and lectures, as one can see from the papers listed below—just a sampling of his seventeen talks and publications in 1914. Included among them is his inaugural lecture to the Prussian Academy of Sciences in Berlin.

Before the family moved to Berlin in April, Einstein had continued his relationship with Elsa by mail, and now, with his presence in Berlin, they became even more involved. Though Mileva had never confronted Albert about the situation, her correspondence and conversations with friends show that she was suspicious and understandably unhappy as he neglected her and left their apartment in the Wilmersdorf section of the city at his pleasure. Finally, in late July, Mileva had had enough emotional turmoil and left Berlin with the boys, returning to Zurich and preparing for an eventual divorce.

In the midst of this personal commotion in Einstein's life, war broke out in Europe that summer, shortly after Mileva and the children had left Germany. After the assassination of Archduke Francis Ferdinand by a Serbian teenager, Gavrilo Princip, Austria and Germany demanded that Serb terrorists be brought to justice, but Serbia and its allies refused. Austria declared war on Serbia, and Germany declared war on Serbia's allies, Russia and France.

As the war spread to other parts of Europe, Germans at all levels of society were expected to support the war effort in some way. Being a Swiss citizen, Einstein was not expected to do so. Indeed, he declared himself a pacifist and now, for the first time, allowed his opinions on nonscientific subjects to be known publicly. He signed a pacifist manifesto drafted by a family friend and physician, Georg Nicolai, which appealed to European rather than nationalist (German) sensibilities. He attended antiwar meetings, one of them held by a pacifist group devoted to establishing a "United States of Europe." The group was subsequently banned in 1916. As history has proven, none of Einstein's appeals worked. He missed his children and worried about them, but he also knew they were safer in neutral Switzerland than in Germany.

A GROUP OF GERMAN astronomers went to Russia to observe the total eclipse of the Sun and test Einstein's new theory of gravity. But World War I erupted on the day of the eclipse, and the German scientists were imprisoned.

Robert H. Goddard, the father of modern rocketry propulsion, began his rocketry experiments in the United

States, though it would be another twelve years before he launched his first rocket.

The Nobel Prize in physics was awarded to Germany's Max von Laue, Einstein's good friend, for having observed that x-rays can be diffracted by solids. Even before this honor, Einstein had considered Laue to be one of Germany's most talented young theoreticians, and Laue was one of a handful of German scientists whom Einstein would still respect after World War II. The Nobel Prize in chemistry went to American Theodore Richards for his accurate determinations of the atomic weight of a large number of chemical elements.

44. "On the Foundations of the Generalized Theory of Relativity and the Theory of Gravitation" (Prinzipielles zur verallgemeinerten Relativitästheorie und Gravitationstheorie). *Physikalische Zeitschrift* 15 (1914): 176–180.

This is a detailed exposition of the hole argument, which implies that the metric tensor $g_{\mu\nu}$ cannot be uniquely determined by generally covariant field equations. This argument is important in reconstructing Einstein's early understanding of the relationship between the coordinate description of a spacetime manifold and its physical properties. Einstein replies to Gustav Mie's criticism of paper 40 (above) by repeating all relevant details and the solution obtained in the earlier paper.

45. "On the Theory of Gravitation" (Zur Theorie der Gravitation). *Naturforschende Gesellschaft in Zurich. Vierteljahrsschrift* 59, part 2, Sitzungsberichte (1914): 4–6.

This is the printed version of a farewell lecture given to a meeting of the Naturforschende Gesellschaft (Society of Natural Scientists) in Zurich on February 9, 1914, shortly before Einstein left for his appointment in Berlin. Einstein reviewed the theoretical aspects of gravitation theory and concluded that there are only two possible versions: Nordström's scalar theory or the Einstein-Grossmann theory.

46. "On the Relativity Problem" (Zum Relativitätsproblem). *Scientia* 15 (1914): 337–348. Einstein made general references to the unsuccessful search for effects of the earth's motion.

47. "The Formal Foundation of the General Theory of Relativity" (Die formale Grundlage der allgemeinen Relativitätstheorie). *Königlich Preussische Akademie der Wissenschaften* (Berlin). Sitzungsberichte (1914): 1030–1085.

According to John Norton ("How Einstein Found His Field Equations"), this major review article was intended to convey the full content of the 1913 "Entwurf" theory (paper 40 above): "The principal novelty lies in the mathematical formulation of the theory. Drawing on earlier work with [Marcel] Grossmann, Einstein formulated his gravitational field equations using a variation principle. Using this richer mathematical structure Einstein offered a proof purporting to demonstrate that his theory had the maximum covariance compatible with the hole argument; that is, covariance under 'justified' transformation between the 'adapted coordinate systems' he had introduced with Grossmann."

48. **"Covariance Properties in the Field Equations of the Theory of Gravitation Based on the Generalized Theory of Relativity" (Kovarianzeigenschaften der Feldgleichungen der auf die verallgemeinerte Relativitätstheorie gegründeten Gravitationstheorie) (with Marcel Grossmann). *Zeitschrift für Mathematik und Physik* 63 (1914): 215–225.**

To increase the inner consistency of the theory expounded in papers 40 and 47 (above), the authors apply an action principle for deriving the field equations for $g_{\mu\nu}$.

49. **"Inaugural Lecture: Principles of Theoretical Physics" (Antrittsrede). *Königlich Preussische Akademie der Wissenschaften* (Berlin). Sitzungsberichte (1914): 732–742.**

In this speech before the Prussian Academy of Sciences, Einstein thanked the members for welcoming him into an academy where he could devote himself exclusively to research (he had no teaching responsibilities here). He also discussed how a theoretician conducts his work, giving relativity theory as an example. Einstein resigned from the academy in 1933 after the rise of the Hitler regime.

50. **"Contributions to Quantum Theory" (Beiträge zur Quantentheorie). *Deutsche Physikalische Gesellschaft. Verhandlungen* 16 (1914): 820–828.**

Einstein attempted to derive Planck's radiation law and Walther Nernst's third law of thermodynamics in a novel manner, based entirely on thermodynamics. The proofs introduced the quantum hypothesis.

51. **"Manifesto to the Europeans" (Aufruf an die Europäer). October 1914. With Georg Nicolai and Wilhelm Foerster.**

In Einstein's first public statement on a nonscientific topic, he responded to a manifesto in which ninety-three German intellectuals and artists had defended German military actions in Europe at the beginning of World War I. This countermanifesto was drafted by pacifist, physician, and professor of medicine and physiology Georg Nicolai and signed by Einstein, Nicolai, Foerster, and Otto Buek. Though it was widely circulated, it was not published until 1917, as the introduction to a book by Nicolai, *The Biology of War*. Ultranationalist Germans considered Nicolai a traitor, and he was barred from teaching in January 1920.

1915

As the Great War continued to plague Europe, Einstein struggled intensively to extend his work on relativity ten years after publishing his first paper on it. By the end of the year, he had a breakthrough and came up with his generalized theory of gravitation. He delivered four lectures to the Prussian Academy of Sciences summarizing his results. His work and his affair with Elsa were now blinding him to the responsibilities of fatherhood, and his sons, particularly Hans Albert, were beginning to resent him and his often-unreasonable demands of them. He wanted to see them in Zurich that summer, but they were away, so he vacationed on the Baltic Sea with Elsa and her daughters. He did keep up his correspondence with the boys and Mileva, however. He visited the boys in September and took Hans Albert on a hiking trip.

In paper 54 (below), Einstein made his feelings on the war known. He was dismayed at the German greed and arrogance displayed after victories over Russia, fearing that a belief in force and conquests would lead only to further harm. As recalled by pacifist Romain Rolland, whom Einstein met in September in Switzerland, Einstein favored splitting Germany into two parts—Prussia in the north and southern Germany and Austria in the south.

ACLU founder Elizabeth Gurley Flynn on trial for inciting strikers to violence

Einstein was lecturing and publishing widely on relativity theory. His 1915 theory replaced the Kepler-Newton theory of planetary motion, which was based on the assumption of absolute space. The new theory was able to account for the slow rotation of the orbital ellipse of a planet. Using Riemann's non-Euclidean geometry and highly nonlinear equations, Einstein was able to predict radically new phenomena—the bending of light around the Sun and the precession of the perihelion of Mercury.

While Einstein was postulating his general theory of relativity, the world became a frightening place as the war swept over Europe. Einstein's friend, chemist Fritz Haber, was put in charge of Germany's chemical warfare and initiated the use of poison gas. Another Einstein

Family displaced by ethnic cleansing in the Ottoman Empire. Some 1.5 million Armenians would be killed or sent into the deserts.

colleague, Walther Nernst, a board member and trustee of the Kaiser Wilhelm Institute, also became a leader in Germany's chemical warfare effort; both of Nernst's sons were killed during the war. Subsequently, the Nobel Prize in chemistry was won by Haber in 1918 and by Nernst in 1920.

EMMY NOETHER, a German mathematician working at the University of Göttingen, proved two mathematical theorems that were basic for both general relativity and elementary particle physics. One is still known as the Noether theorem, proving a relationship between symmetry in physics and conservation principles. Notwithstanding her great abilities and accomplishments, Noether worked without pay as an assistant to David Hilbert because women were not allowed on German faculties. Einstein interceded on her behalf, praising her as a brilliant mathematician, and she was admitted to the faculty—still without pay—in 1919; she finally earned a salary in 1922. At her death, Einstein eulogized that she was "the most significant creative mathematical genius thus far produced since the higher education of women."

Alexander Graham Bell made the first transcontinental phone call from New York to San Francisco, and wireless service was established between the United States and Japan.

The English father-and-son team of Sir William

Henry and Sir William Lawrence Bragg received the Nobel Prize in physics for their analysis of crystal structure by means of x-rays. The chemistry prize went to Germany's Richard Willstätter for his research on plant pigments, especially chlorophyll.

52. "Experimental Proof of Ampère's Molecular Currents" (Experimenteller Nachweis der Ampereschen Molekularströme) (with Wander J. de Haas). *Deutsche Physikalische Gesellschaft. Verhandlungen* 17 (1915): 152–170.

Considering Ampère's hypothesis that magnetism is caused by the microscopic circular motions of electric charges, the authors proposed a design to test Lorentz's theory that the rotating particles are electrons. The aim of the experiment was to measure the torque generated by a reversal of the magnetization of an iron cylinder.

53. "Experimental Proof of the Existence of Ampère's Molecular Currents" (with Wander J. de Haas) (in English). *Koninklijke Akademie van Wetenschappen te Amsterdam. Proceedings* 18 (1915–16).

Einstein wrote three papers with Wander J. de Haas on experimental work they did together on Ampère's molecular currents, known as the Einstein–De Haas effect. He immediately wrote a correction to paper 52 (above) when Dutch physicist H. A. Lorentz pointed out an error. In addition to the two papers above, Einstein and de Haas cowrote a "Comment" on paper 53 later in the year for the same journal. This topic was only indirectly related to Einstein's interests in physics, but, as he wrote to his friend Michele Besso, "In my old age I am developing a passion for experimentation."

54. "My Opinion on the War" (Meine Meinung über den Krieg). Written in October–November 1915 but not published until the following year in *Das Land Goethes 1914–1916*. Stuttgart and Berlin: Deutsche Verlags-Anstalt, 1916.

This statement was written for a volume of "patriotic commemoration," to be published by the Goethebund of Berlin, in which Germans were called on to defend German culture in the midst of war. Einstein declared that war was rooted in the "biologically determined aggressive tendencies of the male." He upheld pacifism and rejected war under any circumstances.

55. "On the General Theory of Relativity" (Zur allgemeinen Relativitätstheorie). *Königlich Preussische Akademie der Wissenschaften* (Berlin). Sitzungsberichte (1915): 778–786. An addendum was added a week later in the same journal.

This important paper on general relativity begins by recounting Einstein's three-year search for the correct field equations of gravitation. He concluded that "the magic of this theory will hardly fail to impose itself on anybody who has truly understood it; it represents a genuine triumph of the method of absolute differential calculus founded by Gauss, Riemann, Christoffel, Ricci, and Levi-Civita."

56. "Explanation of the Perihelion Motion of Mercury from the General Theory of Relativity" (Erklärung der Perihelbewegung des Merkur aus der allgemeinen Relativitätstheorie). *Königlich Preussische Akademie der Wissenschaften* (Berlin). Sitzungsberichte (1915): 831–839.

Einstein gave an astronomical confirmation of general relativity, using as an example the perihelion advance of Mercury.

57. "The Field Equations of Gravitation" (Die Feldgleichungen der Gravitation). *Königlich Preussische Akademie der Wissenschaften* (Berlin). Sitzungsberichte (1915): 844–847.

Dropping a previous restriction, Einstein rewrote his derivations of his field equations of gravitation.

In the three papers above (55–57), all published in November 1915, Einstein explained his new theory of gravitation and the mathematical generalization of the theory of relativity. He concluded that, although spacetime in the special theory is geometrically flat, in the general theory it is curved and includes gravity as a determinant. The seminal paper, however, was not published until 1916 (paper 60 below). Paper 56, however, solved a longstanding problem in astronomy and was one of the major triumphs of the general theory of relativity.

58. "Theoretical Atomism" (Theoretische Atomistic). In *Kultur der Gegenwart*, edited by Paul Hinneberg, vol. 1, *Physik*, part 3, sec. 3. Leipzig: B. G. Teubner, 1915. Also see *The Collected Papers of Albert Einstein*, vol. 4, 521–533. Written in 1913, published in 1915 and again in 1925 in a second edition.

Einstein wrote that an increase in theoretical simplicity and comprehensiveness may be due to a reduced number of fundamental concepts or to independent relations among the concepts. He discussed the quantitative relation among the coefficients of viscosity, heat conduction, and diffusion as established by the kinetic theory of gases.

1916

Einstein became chairman of the German Physical Society, a position he held until 1918. More importantly, his long-awaited details on general relativity were published. From this time on, the 1905 theory became known as *special relativity* and the 1915 theory, which included the phenomenon of gravitation, as *general relativity*. The general theory proposed to explain all laws of physics

Eduard and Hans Albert Einstein in 1916

in terms of mathematical equations. Furthermore, Einstein renewed his work on quantum theory and published three papers on the subject.

During a year in which millions were killed all over Europe and on the seas, Einstein continued his pacifist activities. World War I expanded, with Italy declaring war on Germany and Germany declaring war on Portugal. Food became scarce and was rationed in Germany. Even so, Einstein maintained his friendships with some of those who were heavily invested in the war, such as Fritz Haber and Walther Nernst.

In neutral Switzerland, Mileva's health problems continued to increase, and Einstein became worried about her ability to take care of the boys because she was bedridden and depressed. Indeed, Eduard became weak and ill and was taken to a children's sanatorium in Arosa to recover during the summer. He ended up staying until the following spring.

ARNOLD SOMMERFELD of Germany modified Niels Bohr's atomic model by specifying that electrons have elliptical orbits. Austrian physicist Ernst Mach, who had greatly influenced Einstein's thought, died.

Karl Schwarzschild solved the equations of general relativity in the region outside a spherically symmetric body (a star). His solution shows that the curvature of spacetime becomes infinite if the star's radius is less than $(2GM)/(c^2)$. This is the famous "Schwarzschild radius" of a black hole.

For the first time, the Nobel Prizes in physics and chemistry were not awarded; instead, the prize money was put into the organization's Special Fund for the respective prize sections.

59. "A New Formal Interpretation of Maxwell's Field Equations of Electrodynamics" (Eine neue formale Deutung der Maxwellschen Feldgleichungen der Elektrodynamik). *Königlich Preussische Akademie der Wissenschaften* (Berlin). Sitzungsberichte (1916): 184–188.

60. "The Foundations of the General Theory of Relativity" (Die Grundlage der allgemeinen Relativitätstheorie). *Annalen der Physik* 49 (1916): 769–822.

This long and seminal paper is the earliest complete exposition of Einstein's general theory of relativity, giving a comprehensive account of the final version of the theory after publication of his latest revisions in the papers of November 1915. He presented the tools of tensor analysis that allowed him to derive the general field equations for gravity as a property of non-Euclidean spacetime. In conclusion, he discussed the theory's use in explaining phenomena such as the bending of light rays in a gravitational field, an achievement that helped to establish the superiority of general relativity over classical Newtonian theory and thereby redefined the universe. He wrote an addendum to this paper that was never published.

61. "Approximative Integration of the Field Equations of Gravitation" (Näherungsweise Integration der Feldgleichungen der Gravitation). *Königlich Preussische Akademie der Wissenschaften* (Berlin). Sitzungsberichte (1916): 688–696.

62. "Emission and Absorption of Radiation in Quantum Theory" (Strahlungs-Emission und-Absorption nach der Quantentheorie). *Deutsche Physikalische Gesellschaft. Verhandlungen* 19 (1916): 318–323.

63. "On the Quantum Theory of Radiation" (Zur Quantentheorie der Strahlung). *Physikalische Gesellschaft Zürich. Mitteilungen* 18 (1916): 47–62. Also published the following year in *Physikalische Zeitschrift* 18 (1917): 121–128.

See paper 68 for description.

64. "Elementary Theory of Water Waves and of Flight" (Elementare Theorie der Wasserwellen und des Fluges). *Die Naturwissenschaften* 4 (1916): 509–510.

Einstein had lectured on this topic, in which he proposes an airfoil design, at a meeting of the German Physical Society. On being placed in a moving fluid, a body may experience a net upward force, enabling it to serve as an airfoil. In 1917, a Berlin aircraft firm tested the design, which failed. A few months before his death, Einstein admitted to Paul Ehrhardt, the test pilot, that "I have often been ashamed of my folly of those days."

65. "Ernst Mach." *Physikalische Zeitschrift* 17 (1916): 101–104.

In this long eulogy for Ernst Mach, the natural philosopher who had greatly influenced him, Einstein praised him as an original thinker who had great influence on the epistemological orientation of natural scientists. His "philosophical studies sprang from his desire to find a point of view from which the various branches of science ... can be seen as an integrated endeavor."

1917

As the United States joined the war on the side of the Allies, Einstein wrote his first paper on cosmology, mapping out his first model of the universe. He assumed the cosmos to be spherically symmetric, flat, and static.

Early in the year, Einstein displayed the first symptoms of serious gastrointestinal problems, and within two months he lost fifty-six pounds. It would be four years before he would overcome this affliction. Elsa was only too happy to take care of him, since he was now conveniently situated right across from her fourth-floor apartment in the Schöneberg district of Berlin. By December, he was ill again with an abdominal ulcer and was confined to bed for several months.

In Zurich, Mileva was also ill, as was Eduard. Though hospitals and sanatoriums were generally full in the midst of war, both Mileva and Eduard were able to gain admission to help in their recovery. Einstein wondered whether he should bring Hans Albert to Berlin but was afraid of Mileva's response to the suggestion, so the boy stayed with family friends in Zurich during Mileva's hospitalization. By now, Einstein was sending more than half of his salary to Mileva and the boys and additional funds to his ill mother, so he began to worry about his ability to take care of everyone properly. Though many people in Germany were literally starving and good food was scarce, he was able to get food from his and Elsa's relatives in southern Germany and from his friend Heinrich Zangger in Switzerland. In his letters, Einstein often mentioned the food shortages in Berlin and that, with his already sensitive digestive system, this was a problem.

By the end of the year, Einstein's strained relationship with Hans Albert eased, and his son began to write more affectionate letters.

In October, about three and a half years after Einstein's arrival in Berlin, the Kaiser Wilhelm Institute of Physics finally opened, with Einstein as its first director. Its mission was to promote research in physics and astronomy.

Lenin's address "To the Citizens of Russia" announces the formation of the Soviet Union.

IN VIENNA, as Sigmund Freud wrote his *Introduction to Psychoanalysis,* Britain became committed, through the Balfour Declaration, to establishing a Jewish homeland in Palestine, which had come under its control.

At Mount Wilson in California, a 100-inch reflecting telescope, which Einstein would visit in 1931, was installed.

The Nobel Prize in physics was awarded to Charles Glover Barkla, an Englishman working in Edinburgh, for his discovery of the characteristic Röntgen radiation of the elements. For the second year in a row, the chemistry prize money went into the section's Special Fund.

66. ***On the Special and the General Theory of Relativity: A Popular Account*** **(Über die spezielle und die allgemeine Relativitätstheorie: Gemeinverständlich). Braunschweig, Germany: Vieweg, 1917.**

In this book Einstein presented a popular exposition of relativity theory. Finding it difficult to write, he felt he had no choice but to do so if his theories were to be understood. The book was a huge success, with fourteen editions appearing between 1917 and 1922—and it is still available. Einstein made minor revisions in subsequent editions. Translations into other languages followed, and relativity theory came to be more thoroughly understood throughout the world.

67. **"Cosmological Considerations in the General Theory of Relativity" (Kosmologische Betrachtungen zur allgemeinen Relativitätstheorie).** ***Königlich Preussische Akademie der Wissenschaften*** **(Berlin). Sitzungsberichte (1917): 142–152.**

Einstein's first paper on cosmology was published in the spring. He added a "cosmological constant" in an attempt to balance the field equations of general relativity. It enabled him to describe a universe that conformed to what he and everyone else now assumed: a closed and static sphere that was coextensive with the Milky Way and would neither expand nor collapse on itself; without the cosmological constant, unopposed gravity would cause the universe to collapse. All of modern cosmology goes back to this paper, in which Einstein first applied general relativity to cosmological questions.

Twelve years later, however, astronomer Edwin Hubble, at Mount Wilson Observatory in California, found that the galaxies were speeding away from Earth and that the universe is indeed expanding. He proved what came to be known as the "big bang" theory, which had been proposed by Georges Lemaître, a Belgian Catholic priest and cosmologist who had studied physics at MIT. Einstein, after meeting with Hubble at Mount Wilson Observatory in January 1931 (and also with Lemaître in Pasadena in January 1933), recanted the cosmological constant as the biggest blunder (or words to that effect) of his life and publicly rejected it in paper 171 below. After Einstein heard Lemaître speak about the details of his theory, he is said to have stood up and applauded, saying, "This is the most beautiful and satisfactory explanation of creation to which I have ever listened."

Recent experiments have shown that something is pushing everything farther apart and faster than it did in the early universe, and scientists now believe that the cosmological constant, known by the Greek letter lambda (Λ), is the best candidate to explain it.

---

68. "On the Quantum Theory of Radiation" (Zur Quantentheorie der Strahlung). *Physikalische Zeitschrift* 18 (1917): 121–128.

A penetrating analysis of the properties of photons, this paper also showed that Planck's law for thermal radiation could be simply deduced from assumptions that conform to the basic ideas of the quantum theory of atomic structure, based on the concept of transition probabilities. Einstein formulated general statistical rules regarding the occurrence of radiative transitions among stationary states. In particular, he introduced the possibility of stimulated emission of radiation, which is the principle behind the laser (**l**ight **a**mplification by the **s**timulated **e**mission of **r**adiation).

---

69. "The Nightmare" (Der Angst-Traum). *Berliner Tageblatt*, December 25, 1917.

In this short article, Einstein addressed the question of whether the final exam administered before students could graduate from high school should be abolished. Einstein was in favor of discontinuing the exams, maintaining that they did not succeed in testing a student's knowledge and depended too much on short-term rote learning. He felt that the exam was useless as well as harmful—a teacher could judge students much better over the course of many years than from one final exam, and students had nightmares preceding the examination because their whole future rested on passing it.

1918

The bedridden Einstein, too weak from his stomach problems even to submit his own papers to the various journals, enlisted his colleagues to do so for him. By April he was feeling better and attended some professional meetings. In the fall, he took on his normal teaching load at the University of Berlin.

The Einstein-Marić marriage had no chance of surviving, and in June Einstein signed a divorce agreement with Mileva. Toward the end of the year, the couple took the first formal steps legally to dissolve the marriage. To obtain the divorce, it was necessary for Einstein formally

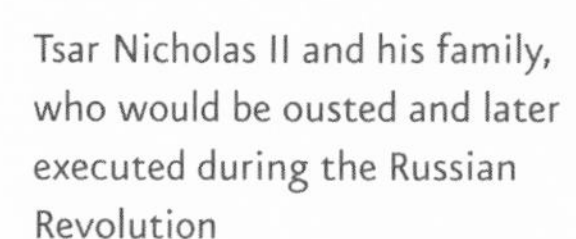

Tsar Nicholas II and his family, who would be ousted and later executed during the Russian Revolution

to declare himself an adulterer and pay a fine for this offense, plus court costs. The court forbade him to remarry for two years, at least in Switzerland.

Just before the divorce was final, Einstein went to Zurich for a month to deliver several lectures and spend time with Mileva and his sons, all of whom had by this time reconciled themselves to the end of the marriage.

The Great War finally ended in November when Kaiser Wilhelm II abdicated in Berlin and an armistice was signed between Germany and the Allies. A great toll of casualties had been taken by both war and illness: an influenza epidemic swept the world, killing 20–40 million before it ended in 1919, while the war dead on both sides numbered 8.5 million. All the Einsteins, even those who were already weakened by illness, survived.

At the dawn of a new regime in Germany, after a new republic was proclaimed, Einstein was optimistic that a free and democratic nation would be possible. He committed himself to left-wing democratic-socialist goals.

HERMANN WEYL in Germany produced, in his paper "Gravitation and Electricity," the first unified field theory in which the electromagnetic and gravitational fields appeared as a property of spacetime. Weyl was one of the first scientists to combine general relativity theory with the laws of electromagnetism. Einstein, however, immediately saw a serious error in the paper, one that would result in the behavior of clocks being dependent on their history. He wrote Weyl on April 15: "If the relation between *ds* and the measuring rod and clock measurements is dropped, the theory of relativity will lose its empirical basis." Weyl fought Einstein on this point for many years, but Einstein prevailed. Even with its problems, the paper was important in ushering in the development of modern gauge theory. Weyl reformulated his work in 1929 after recognizing his errors.

German physicist Max Planck won the Nobel Prize in physics in recognition of his discovery of energy quanta. Planck was Einstein's close personal friend and would profoundly suffer the consequences of two world wars: his elder son was killed in action in 1916, and his daughter died after childbirth the following year; his house was burned down in an air raid during World War II; and his other son was executed in 1945 after admitting his part in a plot to assassinate Hitler. In addition, Planck's first wife had died five years before World War I.

The Nobel Prize committee resumed awarding the prize in chemistry. Einstein's friend Fritz Haber received the prize for synthesizing ammonia from nitrogen and hydrogen, a technique that became known as the Haber process.

70. "On Gravitational Waves" (Über Gravitationswellen). *Königlich Preussische Akademie der Wissenschaften* (Berlin). Sitzungsberichte (1918): 154–167.

The existence of gravitational waves is one of the greatest predictions of general relativity, and it has only recently become technologically possible to search for such waves. Several detectors designed for the purpose were currently under construction.

71. "On the Foundations of the General Theory of Relativity" (Prinzipielles zur allgemeinen Relativitätstheorie). *Annalen der Physik* 55 (1918): 241–244.

72. "Motives for Research" (Motive des Forschens). In *Zur Max Planck's sechzigsten Geburtstag*. Karlsruhe: C. F. Müllersche Hofbuchhandlung, 1918. Reprinted in *Mein Weltbild* (1934) and *Ideas and Opinions* (1954).

This paper was based on a lecture that Einstein delivered in Berlin as part of a special session of the German Physical Society in honor of German physicist Max Planck's sixtieth birthday.

Planck, a professor of theoretical physics at the University of Berlin, is best known for his quantum theory of 1900, the basis for the modern development of atomic physics. Here Einstein described Planck as a model scientist, whose daily efforts on behalf of science come from the heart and not from any other motivations.

73. "The Law of Energy Conservation in the General Theory of Relativity" (Der Energiesatz in der allgemeinen Relativitätstheorie). *Königlich Preussische Akademie der Wissenschaften* (Berlin). Sitzungsberichte (1918): 448–459.

74. "Dialogue on Objections to the Theory of Relativity" (Dialog über Einwände gegen die Relativitätstheorie). *Die Naturwissenschaften* 6 (1918): 697–702.

This article on the paradoxes of the theory of relativity was written as a dialogue between Einstein and his critics.

75. "On the Need for a National Assembly." November 13, 1918. Published in Nathan and Norden, *Einstein on Peace.*

This speech, given by Einstein at a public meeting of the German nonpolitical but pacifist organization Bund "Neues Vaterland," advocated for a legislative assembly. The BNV was devoted to promoting individual self-development through a genuine intellectual and moral culture and favored political parties that would provide such leadership. Around this time, Einstein began to defend himself against unfounded accusations that he was an anarchist or communist.

76. "To the Society 'A Guaranteed Subsistence for All'" (An den Verein 'Allgemeine Nährpflicht'). December 12, 1918. Unpublished, but see *The Collected Papers of Albert Einstein*, vol. 7, doc. 16 and its footnotes.

In this one-page statement, Einstein discussed the program of Austrian philosopher and inventor Josef Popper-Lynkeus to guarantee every person a minimum of goods and services needed for subsistence, especially in times of war and unemployment. Einstein said this was the most important social goal of the time, though he disagreed with how Popper planned to run the program. Popper's views were, in general, unpopular because his program put the individual's needs ahead of those of the state or of science.

1919

The year 1919 was a watershed in Einstein's personal life: he divorced Mileva; he married Elsa, who would remain his wife until her death seventeen years later; he became an international celebrity; and he began to identify more and more with his Jewish brethren.

President Theodore Roosevelt, here seated in his rocking chair, dies in 1919.

Beginning the year tumultuously, Berlin experienced the left-wing and ill-organized "Spartacist uprising" early in January and the murder of its Communist leaders Rosa Luxembourg and Karl Liebknecht on January 15. Einstein was in Zurich at the time, getting ready to deliver his first cycle of lectures as a visitor at the University of Zurich.

Early in the year, Einstein moved across the hall into Elsa's apartment, where he was given his own room and study. In February, ironically on Valentine's Day, he and Mileva were formally divorced, with the divorce decree stipulating that any future Nobel Prize moneys would go to Mileva to support the boys. Albert and Elsa were thus free to marry in June. The obligatory two-year waiting period set by the Swiss judge was not enforceable in Germany. By now, Ilse was twenty-two years old and Margot was twenty, and both young women continued to live with the Einsteins, along with Einstein's mother's sister, Fanny (who was also Elsa's mother), and brother, Jakob. Only one year earlier Ilse had confided to family friend Georg Nicolai, a physician and pacifist, that Einstein had proposed marriage to her—with her mother's knowledge—but that she had turned him down, saying that her love for him was more like that for a father than for a prospective spouse.

Just four days before the wedding on June 2, a British scientific expedition to the island of Principe off the West African coast, led by Sir Arthur Eddington, England's foremost expert on Einstein's theory of general relativity and director of the Cambridge Observatory, measured the bending of starlight by the gravitational field of the Sun. A photograph of the total solar eclipse bore out Einstein's theory that gravity would bend light. Einstein became a world celebrity after the Royal Society announced the findings in November. Because the theory was so difficult to understand and explain in nonscientific terms, it became all the more mysterious, adding to the myth that only Einstein could see a dimension to the universe that was beyond almost anyone else's comprehension. He enjoyed his sudden fame and was jovial with reporters; eventually, however, he began to retreat from the commotion of lifelong celebrity.

Einstein devoted much of his time now to problems of international reconciliation. Through his friendship with Zionist activist Kurt Blumenfeld, he became interested in Zionism and began to support the idea of a Jewish state in Palestine. Until he had come to Berlin, his Jewishness had never been an issue with him, for he was totally neutral about ethnic distinctions and felt at home anywhere he went. Although he supported the Zionist ideal of a Jewish homeland, he never joined any Zionist organization and did not consider himself a Zionist. In a letter of March 25, 1955, Einstein thanked Blumenfeld for having "helped me become aware of my Jewish soul."

ERNEST RUTHERFORD, still working on the structure of the atom, discovered the proton, the positively charged particle within the nucleus. He published the first evidence of artificially splitting atomic nuclei by producing hydrogen through the bombardment of nitrogen with alpha particles. His discovery made possible the description of electrostatic force: if each of two bodies has an excess of electrons or protons, repulsion occurs, but if the two bodies differ in their excesses, attraction occurs.

The Boxer Rebellion in China, begun in 1900, leads to the trial of rebel leaders.

Francis Aston of England built the first mass spectrograph capable of measuring the masses of the elements with useful precision.

The Nobel Prize in physics was awarded to Johannes Stark of Germany for his discovery of the Doppler effect in canal rays and the splitting of spectral lines in electric fields. Canal rays, discovered in 1886, are positive rays that played an important role in the development of early-twentieth-century physics but are not widely known today. The chemistry prize money again went into the section's Special Fund.

77. "Do Gravitational Fields Play an Essential Role in the Structure of the Elementary Particles of Matter?" (Spielen die Gravitationsfelder im Aufbau der materiellen Elementarteilchen eine wesentliche Rolle?) *Königlich Preussische Akademie der Wissenschaften* (Berlin). Sitzungsberichte (1919): 349–356.

In this talk to the Prussian Academy of Sciences, Einstein visualized the relations in spherical space and discussed the field equations in general relativity from the point of view of the cosmological problem and the problem of the constitution of matter.

78. "Comment on Periodical Fluctuations of Lunar Longitude, Which So Far Appeared to Be Inexplicable in Newtonian Mechanics" (Bemerkung über periodische Schwankungen der Mondlänge, welche bisher nach der Newtonschen Mechanik nicht erklärbar schein-en). *Königlich Preussische Akademie der Wissenschaften* (Berlin). Sitzungsberichte (1919): 433–436.

79. "A Test for the General Theory of Relativity" (Prüfung der allgemeinen Relativitätstheorie). *Die Naturwissenschaften* 7 (1919): 776.

In one small paragraph, Einstein commented on Eddington's expedition to the island of Principe to test the general theory of relativity and reported the results.

80. "Time, Space, and Gravitation." *Times* (London), November 28, 1919.

Writing at the request of the *Times*, Einstein expressed his appreciation to England and English scientists for sparing no expense in testing his theory, "which was perfected and published during the war in the land of your enemies." He also discussed various theories and principles in physics as they apply to the theory of relativity.

81. "Induction and Deduction in Physics" (Induktion und Deduktion der Physik). *Berliner Tageblatt*, December 25, 1919, morning edition, sec. 4.

This article was part of a series of articles by prominent German scientists on the achievements of German science. Einstein warned that "the truth of a theory can never be proven, for one never knows if future experience will contradict its conclusions."

82. "Immigration from the East" (Die Zuwanderung aus dem Osten). *Berliner Tageblatt*, December 30, 1919, morning edition.

Einstein spoke out against those who claimed that Jews coming in from eastern Europe were responsible for Germany's problems after the war.

83. "Leo Arons as Physicist" (Leo Arons als Physiker). *Sozialistische Monatshefte* 52, part 2 (1919): 1055–1056.

An obituary for a physicist whom Einstein admired for his political courage.

1920

In February, Einstein's mother, who had come to live with him and Elsa in Berlin because of her illness, died at the age of sixty-two. He wrote to his friend Heinrich Zangger that "my mother died a week ago today in terrible agony. We are all completely exhausted."

Anti-Jewish sentiment began to be seen in Berlin, especially directed against Einstein, the new celebrity. Outrageous criticisms of relativity theory—which anti-Semites dubbed the "Jewish science"—came particularly from German physicists Philipp Lenard and Johannes Stark, both of whom had won the Nobel Prize.

Einstein tried to ignore these sentiments and involved himself with postwar reconstruction. Distraught at the high child mortality and mass hunger he saw around Berlin, he thanked American and British Quakers for their efforts in feeding more than a half-million German children. By the end of the year, the League of Nations was ready to assume a major role in humanitarian relief efforts.

Einstein made lecture trips to Holland, where he was given an appointment as visiting professor; to Norway; and to Denmark, where he visited physicist Niels Bohr, whom he had already met during Bohr's trip to Berlin in February.

Adolf Hitler, at the Hofbräuhaus in Munich, pro-

-Semitism because many unregistered "students" in the audience appeared to be Jews from ern Europe who had immigrated to Berlin, but the student council, administration, and tein himself denied this. As a result of this "uproar," which continued for several days, the inistration changed its admissions policy. Soon after this event, Einstein reversed his stand tly by saying that the unregistered public should be allowed into the lecture hall if seats still available after the paying students had been seated. Just in case all seats were taken, so arranged to give a series of free evening lectures outside the elite university.

ti-Semitism was growing in Germany after the war, and Einstein wrote two documents on Jewish question" around this time, but they were not published.

"A Confession" (newspaper report quoting a statement by Einstein). *Israelitisches Wochenblatt für die Schweiz*, September 24, 1920.

nstein had been invited to take part in a meeting dedicated to fighting anti-Semitism ademic circles, but he declined because he thought such an endeavor would not prove ul. In this statement, he maintained that Jews must eliminate the anti-Semitism among selves (western European Jews against eastern European Jews) before they could earn spect of others. "Only when we dare to see ourselves as a nation, only when we respect ves, can we earn the respect of others, or rather, can they arrive at this conclusion them-." He was not concerned with the anti-Semitism exhibited by non-Jews—here he states sly that perhaps it is due to such prejudice that Jews have historically banded together rvived as a race. "Anti-Semitism will be a psychological phenomenon so long as Jews come ontact with non-Jews—what harm can there be in that? . . . I am a Jew, and I am glad to to the Jewish people, even though in no way do I consider them the Chosen Ones . . . My sion is not meant to be malicious or unfriendly!"

***Ether and the Theory of Relativity* (Äther und Relativitätstheorie). Berlin: Springer, 1920.**

tein's lecture at the University of Leiden on the occasion of his appointment as a visiting sor summarized his current views on the ether and retrospectively looked at the develop-f his opinions on the physical properties of space.

"Propagation of Sound in Partly Dissociated Gases" (Schallausbreitung in teilweise dissoziierten Gasen). *Königlich Preussische Akademie der Wissenschaften* (Berlin). Sitz-richte (1920): 380–385.

paper is off the beaten track for Einstein, who dabbled with fluids (liquids and gases) nout his life.

"To the 'General Association for Popular Technical Education'" (newspaper report quoting a statement by Einstein). *Neue Freie Presse*, July 24, 1920, morning edition.

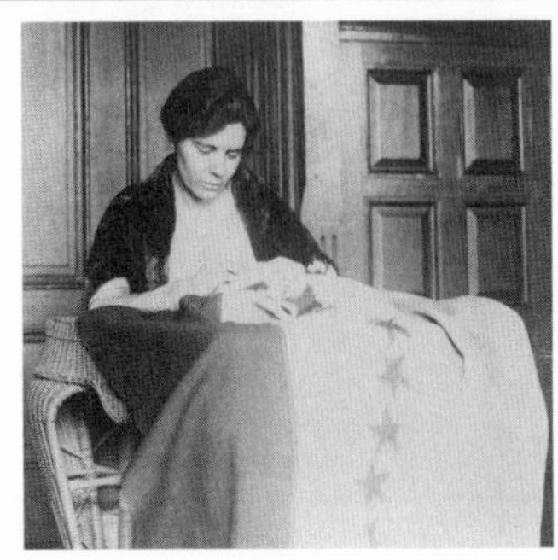

Alice Pavl sews the thirty-sixth star on the suffrage ratification banner. Women finally get to vote in 1920.

nounced his twenty-five-point progra reform in Germany. The German W renamed the National Socialist Germ Nazi—Party, ushering in an era of fasci would last twenty-five years.

Chaim Weizmann, a Polish-born whom Einstein later befriended, bec World Zionist Organization. Weizma the first president of Israel in 1948.

IN PHYSICS, Ernest Rutherford postu stating that it was required to keep the protons inside the nucleus from rep The existence of neutrons explains wh the same chemical properties but sligh

The Nobel Prize in physics was a Edouard Guillaume of Switzerland measurements in physics by which h alies in nickel-steel alloys. The prize to Walther Nernst of Germany for h chemistry.

84. "Uproar in the Lecture Hall" (newspaper account containing a q Einstein). *8-Uhr Abendblatt*, February 13, 1920.

Einstein had a policy of allowing students who were not registered a tend his lectures. The student council protested his open-admission p paying students and heckled him during a lecture. Some claimed that t

The association had solicited this statement. Einstein stressed the importance of showing to students and the public the practical applications of science and technology to everyday life, stating his belief that a technical education is equal in value to a humanistic one.

89. "On New Sources of Energy" (newspaper report quoting a statement by Einstein). *Berliner Tageblatt*, July 25, 1920, morning edition.

The French were demanding coal from the Germans as war reparations. Einstein postulated that atomic energy might be a powerful alternative source, though he had no data yet to support this idea.

90. "My Response. On the Anti-Relativity Company" (Meine Antwort. Über die anti-relativitätstheoretische G.m.b.H.). *Berliner Tageblatt*, August 27, 1920, morning edition.

In the popular Berlin newspaper, Einstein repudiated the attacks against him and relativity theory since 1918. The final straw had been two lectures he delivered on August 24 in the Berlin Philharmonic Hall, where his accusers put aside any scientific criticisms and charged him with plagiarizing, being hungry for publicity, manipulating the press, conducting un-German science, and so on. Outside the hall, his enemies distributed anti-Semitic leaflets and sold swastika lapel pins. Einstein responded to them angrily in this statement and in a lecture a month later, after having resisted public responses to earlier provocations. These encounters polarized the German physics community, and Einstein's supporters were afraid the campaign against him might drive him out of Germany.

91. "On the Contribution of Intellectuals to International Reconciliation." In *Thoughts on Reconciliation*. New York: Deutscher Gesellig-Wissenschaftlicher Verein, 1920.

The German Social and Scientific Society of New York solicited this statement for an anniversary volume published in the fiftieth year of its founding. The proceeds from the sale were to help needy intellectuals in Germany and Austria.

92. "Interview on Interplanetary Communication." *Daily Mail* (London), January 31, 1920. An indirect quotation is in *New York Times*, February 2, 1920.

Prompted by "mysterious wireless signals" from an unknown source received in both London and New York, a London correspondent contacted Einstein for an explanation. Einstein said that there is every reason to believe that Mars and other planets are inhabited but that Martians would be more likely to communicate via light rays than through the wireless. He hypothesized that the signals were due to either atmospheric disturbances or secret experimentation with other systems of wireless telegraphy.

1921

Einstein in a New York motorcade

In the spring, Einstein undertook a double mission on his first trip to America: a fund-raising tour on behalf of the Hebrew University of Jerusalem and a visit to Princeton University in New Jersey to deliver a series of four lectures on relativity theory. Part of his motivation for the Princeton lectures was financial—he was strapped for money because of his many family obligations, and he was looking for financial resources from abroad during the difficult postwar economic situation in Germany. He insisted that Elsa accompany him because of his poor health. Leading the trip on behalf of the Zionists and Hebrew University was future Israeli president Chaim Weizmann, the new president of the World Zionist Organization and a British citizen.

Upon his arrival in New York, Einstein immediately became popular as he joked with reporters and as city officials took him on a ticker-tape motorcade through the city. In April, he gave lectures to packed auditoriums at Columbia University and the City University of New

York. Before heading to Princeton, he delivered a lecture to the National Academy of Science in Washington, D.C., and briefly visited the White House to meet President Warren Harding. His reception in Washington was fairly cool because of his connection with Germany, as Americans had not warmed up to the postwar German government.

In Princeton, Einstein was awarded an honorary doctorate from the university. He also delivered four lectures, published as a book the following year both in England and in the United States; it is still in print. The fundraising part of the trip was not as successful as the Zionists had hoped, but during this time Einstein seemed to cement his Jewish identity.

Before returning to Germany, the Einsteins stopped in England, where Einstein gave lectures at King's College at the University of London (published in 1934 in *The World as I See It*) and at Manchester University. He also visited the tomb of Newton, leaving a bouquet of flowers. Later in the summer, he took a vacation to go sailing with his two sons.

Though Einstein did not receive the news until the following year, he was awarded the 1921 Nobel Prize in physics for his 1905 work on the photoelectric effect. Because some physicists were still disputing relativity theory, the Nobel award did not mention it.

In Berlin, Einstein met physicist and inventor Leo Szilard, who would later became his collaborator and convince him to sign the famous letter to President

Eddie Cicotte throws a pitch before being banned from the game for life after his involvement in the Black Sox scandal.

Einstein with Paul Ehrenfest and his son

Franklin Roosevelt alerting him that the Germans might be building an atomic bomb.

Hitler's stormtroopers began to terrorize his political opponents as the German mark plummeted and a steep inflationary period set in.

IN PHYSICS, Theodor Kaluza wrote down Einstein's field equations in five dimensions, reproducing the usual four-dimensional gravitational equations plus Maxwell's equations for the electromagnetic field. According to this hypothesis, electromagnetism is not a separate force but an aspect of gravity in a higher dimension.

The physics prize went to Einstein, and the Nobel Prize in chemistry was awarded to Frederick Soddy of England for his work on the chemistry of radioactive substances and his investigations into the origin and nature of isotopes.

93. "Einstein on Education." *Nation and Athenaeum* 30 (1921): 378–379. This article was not written by Einstein but was probably based on an interview with him, quoting him on matters of mathematical and scientific education. Einstein concluded that present-day education is too concerned with abstractions and should be more hands-on, that testing students with exams should not be necessary, and that the school day should be a maximum of six hours.

94. "The Common Element in Artistic and Scientific Experience" (Das Gemeinsame am künstlerischen und wissenschaftlichen Erleben). *Menschen. Zeitschrift neuer Kunst* 4 (1921): 19.

In this short statement, Einstein emphasized that art and science have common elements. Both give expression to creativity: in science through logic and in art through form, and both help one escape from merely personal concerns.

95. *Geometry and Experience* (Geometrie und Erfahurung) (the last part of a lecture, which appeared first as a reprint). Berlin: Springer, 1921. (English version published in *Sidelights on Relativity*. London, Methuen, 1922.)

Lecturing to a commemorative session of the Prussian Academy of Sciences in honor of Frederick the Great, Einstein discussed the special esteem enjoyed by mathematics. He summed up his views on the geometrization of physics and relativity and the relation of mathematics to the external world. He questioned whether human reasoning, even without direct experience, could lead to an understanding of the properties of real things merely through thought. Considering the puzzling question of why mathematics should be so well adapted to describing the external world, he concluded: "As far as the laws of mathematics refer to reality, they are not certain; and as far as they are certain, they do not refer to reality."

96. "A Brief Outline of the Development of the Theory of Relativity." *Nature* 106 (1920–21): 782–784. Translation by Robert Lawson based on a German manuscript.

In a special issue of *Nature* devoted to relativity theory, Einstein and other European contributors wrote about the contemporary results and problems of the theory in the hope of restoring international scientific cooperation after the war. Einstein described the sequence of ideas that had led to his theory and concluded with prescient remarks on the remaining questions.

97. "On a Natural Addition to the Foundation of the General Theory of Relativity" (Über eine naheliegende Ergänzung des Fundamentes der allgemeinen Relativitätstheorie). *Königlich Preussische Akademie der Wissenschaften* (Berlin). Sitzungsberichte (1921): 261–264.

98. "In My Defense" (Zur Abwehr). *Die Naturwissenschaften* 9 (1921): 219.

Einstein expressed outrage about a preface fabricated by Lucien Fabre and attributed to Einstein, published within Fabre's popular article on relativity for a French journal. Fabre was a French engineer, poet, and novelist who had claimed in another article that Henri Poincaré was the true discoverer of relativity. Fabre had a reputation for being a "reactionary and Jew-hater."

99. "A Simple Application of the Newtonian Law of Gravitation to Globular Star Clusters" (Eine einfache Anwendung des Newtonschen Gravitationsgesetzes auf die

kugelförmigen Sternhaufen). In *Festschrift der Kaiser-Wilhelm-Gesellschaft zur Förderung der Wissenschaften zu ihrem zehnjährigen Jubiläum*, 50–52. Berlin: Springer, 1921.

A contribution to the *Festschrift* published on the occasion of the tenth anniversary of the Kaiser Wilhelm Society of Berlin.

100. **"How I Became a Zionist" (Wie ich Zionist Wurde). *Jüdische Rundschau*, June 21, 1921, pp. 351–352.**

In an interview by the editor of the Jewish magazine, Einstein maintained that intellectuals are the propagators of anti-Semitism, that elites use anti-Semitic sentiments for political gain, but that hatred comes from those who are "primitive and uneducated" about those who are different from them. He reiterated his indifference to Judaism as a religion.

101. **"On a Jewish Palestine" (based on an address given by Einstein in Berlin on June 27, 1921). *Jüdische Rundschau*, July 1, 1921, p. 371.**

Einstein emphasized the importance of Palestine as a symbol of cultural unity over its importance as a Jewish settlement; cultural unity was the basis for his support for Zionism at this time. In the English-language translation of the preceding article in the *Jewish Chronicle* of June 17, 1921, Einstein had spoken out against confining Jewish ethnic nationalism to Palestine, believing that Jews should be acknowledged in the Diaspora wherever they lived; otherwise the world would deny the existence of a Jewish people. He felt that American Jews were more aware of this fact than were German Jews, who seemed more eager to assimilate.

102. **"Einstein's Impressions of America: What He Really Saw" (Einsteins amerikanische Eindrücke. Was er wirklich sah). *Vossische Zeitung*, July 10, 1921, morning edition, suppl.**

After Einstein's trip to America in 1921, he made some remarks about Americans that were not popular back in the States, especially one in which he characterized American men as the lap dogs of their wives. In this article he attempted to explain, if not recant, his statements. He did say that his previously published remarks were misrepresented. He commended Americans on their warmth and friendliness and expressed admiration for their close relationships between students and teachers. He also said he felt that American patriotism is not so much nationalistic as an inner pride of country.

103. **"On the Founding of the Hebrew University of Jerusalem" (Zur Errichtung der hebräischen Universität in Jerusalem). *Jüdische Pressezentrale Zürich*, August 26, 1921.**

In this article, Einstein supported the establishment of a university in Jerusalem, advocating an emphasis on the sciences and health professions. He felt that it was necessary for a Jewish homeland to have a university both for its own citizens and also to give Jews who had been barred from other universities a place to study, teach, and conduct research. The hope was that, when such an institution gained a distinguished international reputation, Jews would no longer be inclined to hide their group identity wherever they might work.

104. "On an Experiment Concerning the Elementary Process of Light Emission" (Über ein den Elementarprozess der Lichtemission betreffendes Experiment). *Königlich Preussische Akademie der Wissenschaften* (Berlin). Sitzungsberichte (1921): 882–883.

105. "The Plight of German Science: A Danger for the Nation" (Die Not der deutschen Wissenschaft. Ein Gefahr für die Nation). *Neue Freie Presse*, December 25, 1921, morning edition.

A statement solicited by the editor to promote scientific interests between Germany and Austria.

1922

As Wilhelm Cuno took over as chancellor of Germany, Einstein completed his first paper on a unified field theory, for which he would continue to search for the rest of his life. In his hometown of Ulm, a small street was named after him; in 1933 the Nazis changed the name to something else, but it was restored after the war.

With Marie Curie, H. A. Lorentz, and others, Einstein joined the League of Nations' Commission on Intellectual Cooperation.

Einstein's distractions became more and more nonscientific as the times hurled him, along with millions of others, into the politics and social climate of a changing world. Germany was continuing to experience economic hardship and depression, communism was gaining adherents in the east, being especially popular among the poor, and many thought Jews were good scapegoats for both these circumstances. Anti-Semitism became widespread.

Still, most Germans celebrated Einstein. They wanted the prestige of his name to add gloss to their institutions even as he criticized them. He remained friendly, informal, and "democratic" toward all, believing that one should respect honest people even when they have views different from one's own. After a nationalist fanatic as-

Physicist Niels Bohr

sassinated the newly elected German foreign minister, Walther Rathenau, a Jew and a friend, Einstein took heed and began to remove himself from any controversial political arenas. Indeed, he was warned not to make any public appearances for the sake of his safety. He wrote to his friend Maurice Solovine on July 16: "There has been much excitement here ever since the abominable murder of Rathenau. I myself am being constantly warned to be cautious, have canceled my lectures, and am officially 'absent,' although actually I have not left." But he did begin to think about resigning from the Kaiser Wilhelm Institute and moving away.

First, however, came the opportunity to take another extended trip abroad, this time to Japan, with stops in Palestine and Spain on the return trip. He and Elsa were the guests of a Japanese publishing house, Kaizosha, which offered him generous compensation for a series of lectures throughout Japan. The Einsteins left Germany at the beginning of October and headed to Marseilles to board the Japanese steamer *Kitanu Maru.* Stops along the way included Singapore, Hong Kong, Ceylon, and Shanghai.

Einstein again developed abdominal pains and was attended by the ship's doctor. On his way to Shanghai, he received news that no doubt made him feel much better: he had been awarded the Nobel Prize in physics for 1921 for his work on the photoelectric effect, published in 1905. With the added prestige of being a Nobel laureate, he lectured in Tokyo, Sendai, Kyoto, and Fukuoka during his six-week tour of Japan, which he ended by playing his violin at a YMCA Christmas party.

A month and many memories later—which he recorded in a travel diary that is still in the Einstein archive (the portion about his trip to Japan has been published in that country)—he was on the shores of Palestine. His visit would leave a lasting impression on him and influence the direction of his future life.

EDWIN HUBBLE, a former lawyer who became an astronomer, demonstrated from 1922 to 1924 that nebulae containing Cepheid stars are not located within the Milky Way Galaxy but are situated in other, more distant galaxies. This discovery revolutionized our knowledge about the cosmos. Russian physicist Alexander Friedmann, based on earlier work by Willem de Sitter, showed that the universe—as described by Einstein's equations—could expand or contract.

Niels Bohr of Denmark won the Nobel Prize in physics for his investigations into the structure of atoms and the radiation emanating from them. Francis W. Aston of England, a keen sportsman and musician as well as scientist, received the award in chemistry for the discovery, by means of his mass spectrograph, of isotopes in a large number of nonradioactive elements and for his enunciation of the whole-number rule.

106. "Conditions in Germany." *New Republic* 32 (1922): 197.

107. "In Memorium Walther Rathenau." *Neue Rundschau* 33, part 2 (1922): 815–816. ■ A eulogy for the slain German foreign minister, assassinated by two right-wing army officers on June 24, 1922. Leaders of the Nazi Party had accused Rathenau of being part of a "Jewish-Communist conspiracy."

108. "Impact of Science on the Development of Pacifism." German contribution to Kurt Lenz and Walter Fabian, eds., *Die Friedensbewegung* (The Peace Movement), 78–79. Berlin: Schwetschke, 1922.

Because technical inventions that arise from science have international consequences, including military applications, men must create organizations dedicated to preventing wars whenever there is a possibility that such products might be used violently.

109. "Four Lectures on the Theory of Relativity, Held at Princeton University in May 1921" (*Vier Vorlesungen über Relativitätstheorie gehalten im Mai 1921 an der Universität Princeton*). Braunschweig, Germany: Vieweg, 1922. (English version published by Methuen in London as *The Meaning of Relativity*, translated by Edwin P. Adams. See also 5th ed., pbk., Princeton Science Library. Princeton, N.J.: Princeton University Press, 1988.)

In 1921, Einstein had made his first trip to the United States, chiefly on a mission to raise funds for the establishment of the Hebrew University of Jerusalem. Princeton University had invited him to give a series of three lectures a week over a two-month period, for which he asked fifteen thousand dollars in compensation. In a compromise, the series was cut to four lectures called "The Meaning of Relativity," for which he received a much smaller fee. This book is based on these lectures.

110. "Theoretical Observations on the Superconductivity of Metals" (Theoretische Bemerkungen zur Supraleitung der Metalle). In *Gedenkboek Kamerlingh Onnes* (Commemorative Volume for Kamerlingh Onnes), 429–435. Leiden: Ijdo, 1922.

111. "Experiment Concerning the Limits of Validity of the Wave Theory" (Experiment betreffend die Gültigkeitsgrenze der Undulationstheorie). *Königlich Preussische Akademie der Wissenschaften* (Berlin). Sitzungsberichte (1922): 4.

112. "Theory of the Propagation of Light in Dispersive Media" (Theorie der Lichtfortpflanzung in dispergierenden Medien). *Königlich Preussische Akademie der Wissenschaften* (Berlin). Sitzungsberichte (1922): 18–22.

113. "Quantum-Theoretical Observation on the Stern-Gerlach Experiment" (Quantentheoretische Bemerkung zum Experiment von Stern und Gerlach) (with Paul Ehrenfest). *Zeitschrift für Physik* 11 (1922): 31–34.

The two physicists demonstrated that the Stern-Gerlach effect, discovered in 1922, had to overcome insurmountable difficulties if it were to show the behavior of atoms in a magnetic field. See paper 198 for more on Ehrenfest, one of Einstein's closest friends.

114. "How I Discovered the Theory of Relativity" (Wie ich die Relativitätstheorie entdeckte). Informal lecture at the University of Kyoto, Japan, December 14, 1922. (Notes taken by Yun Ishiwara and translated into English by Y. A. Ono in *Physics Today,* August 1932, p. 45.)

Einstein's translator on this trip was Japanese physicist Yun Ishiwara. Ishiwara published his notes of the lecture in 1923 in Tokyo and had them reprinted there in 1971.

1923

Early in the year, Einstein and his wife, Elsa, were completing their journey to Palestine, where Einstein laid the cornerstone of the first building of the Hebrew University of Jerusalem. The couple traveled on to France and Spain before returning to Berlin in the middle of March.

In July, Einstein went to Sweden to deliver his Nobel lecture. After some disagreements with Mileva about how to use the Nobel Prize money, they decided to buy three multifamily houses in the Zurich area, whose rental income should assure Mileva's financial security. That summer, Einstein spent time with his teenage sons vacationing in southern Germany.

Einstein resigned from his post on the League of Nations' Commission on Intellectual Cooperation because he felt it had neither the strength nor the will to accomplish its stated mission, and as a pacifist he could no longer support it. He tried to come to terms with his Prus-

Crossing from the Atlantic to the Pacific Ocean, or back, is easy now that the Panama Canal is a fully operational success.

Booze is banned. Prohibition officers raid a D.C. speakeasy.

sian and German nationality as a period of economic stability seemed to begin with the end of inflation.

Hitler's "Beer Hall Putsch" (coup d'état) in Munich failed, and he was arrested and subsequently sentenced to five years in prison. Fascism was also sweeping Italy, whose leaders were dissolving all nonfascist political parties.

SIGMUND FREUD—with whom Einstein would correspond in 1929, 1932, and 1936—published *The Ego and the Id,* and Arthur Eddington published *The Mathematical Theory of Relativity,* which Einstein considered the finest presentation of the subject in any language.

Sigmund Freud's psychoanalysis revolutionizes psychology.

The Electrolux Company produced the first electric refrigerator, and Vladimir Zworykin, a Russian immigrant working at Westinghouse, invented and patented the first iconoscopic television camera tube. It was not until 1929, however, after he teamed up with another Russian immigrant, David Sarnoff, at RCA that he received the managerial and financial backing to develop his invention into a marketable product. The RCA team finally introduced its television ten years later at the 1939 World's Fair in New York.

Robert Millikan of the United States was awarded the Nobel Prize in physics for his experimental work that determined the charge on the electron and on the photoelectric effect. Millikan had met Einstein at Caltech in Pasadena two years earlier. The chemistry prize went to

Fritz Pregl of Austria for his method of microanalyzing organic substances.

---

115. "My Impressions of Palestine." *New Palestine* 4 (1923): 341.
Einstein recounted his experiences and feelings during his visit to the future Jewish homeland. He was not yet concerned about an "Arab question," declaring that Jews and Arabs appeared to live in harmony and that the major problems were those of sanitation, malaria, and debt. By the end of the decade, after major anti-Jewish disturbances, he became more concerned about the Arabs and their rights.

---

116. "On the General Theory of Relativity" (Zur allgemeinen Relativitätstheorie). *Königlich Preussische Akademie der Wissenschaften* (Berlin). Sitzungsberichte (1923): 32–38 (note appended on 76–77).

---

117. "Theory of Relativity." *Nature* 112 (1923): 253. Translation of an article written first in French for the *Bulletin of the Société française de philosophie* 22 (1923): 97–112.

---

118. "Theory of the Affine Field." *Nature* 112 (1923): 448–449.

---

119. "Fundamental Ideas and Problems of the Theory of Relativity" (Grundgedanken und Probleme der Relativitätstheorie). In *Nobelstiftelsen: Les prix Nobel en 1921–1922*. Stockholm: Imprimerie Royale, 1923.

This paper presents the Nobel lecture delivered in Göteburg, Sweden, on July 11, 1923. Because Einstein did not deliver the lecture at the same time as he received the award, it did not concern the prize topic—the discovery of the photoelectric effect—but surveyed relativity theory instead.

---

120. "Does Field Theory Offer Possibilities for the Solution of Quantum Problems?" (Bietet die Feldtheorie Möglichkeiten für die Lösung des Quantenproblems?). *Königlich Preussische Akademie der Wissenschaften* (Berlin). Sitzungsberichte (1923): 359–364.

Despite great successes in quantum theory for a quarter of a century, there was still no logical foundation for the theory, and one must question whether the remaining difficulties could be overcome by the consistent development of earlier theories.

---

121. "On the Quantum Theory of the Radiative Equilibrium" (Zur Quantentheorie des Strahlungsgleichgewichts) (with Paul Ehrenfest). *Zeitschrift für Physik* 19 (1923): 301–306.

1924

The Einsteins celebrated the wedding of Elsa's elder daughter, Ilse, who married journalist Rudolf Kayser. In 1930 Kayser would publish the first biography of Einstein under the pen name of Anton Reiser. After the marriage of Ilse, whom he had once loved, Einstein became enamored with his new secretary, Betty Neumann, but ended the relationship by the end of the year. Later, a succession of other women entered Einstein's life. He had affairs with some and ignored others, especially those who wrote to him offering to become his wife after he became a widower. His marital relationship with the good-natured Elsa was grounded on companionship rather than passion, and there is no indication that she was jealous about her husband's dalliances. More than likely, she resigned herself to these short-term affairs and took pleasure in being the wife of a famous man, confident that he would never leave her.

Having second thoughts about giving up his membership, Einstein rejoined the League of Nations commission he had abandoned the previous year. His goal now was to use the organization to effect worldwide standardization of terminology in physics and chemistry. He even attended a meeting of the commission in Geneva. In December, the Einstein Tower observatory and laboratory, an expressionist, early modern piece of architecture, was officially opened in Potsdam near Berlin.

Marcus Garvey founds the Back-to-Africa movement.

By the middle of May, Einstein wrote to a friend that political conditions in Germany had become more settled and that his life was more tranquil. Although Hitler was sentenced to five years in prison for his attempted coup, he was released after eight months for "good behavior." While in prison, he wrote *Mein Kampf* (My Struggle), in which he expressed his hatred of Jews and Slavs, berated democracy, and detailed his blueprint for achieving German supremacy.

In December, Einstein made his last major scientific discovery: by analyzing statistical fluctuations he arrived at an independent argument for the association of

Babe Ruth, "The Sultan of Swat," crosses home plate for the New York Yankees.

waves with matter. He also predicted that a state of matter might exist in which a collection of bosons, below a certain temperature, all occupy the ground-state energy level. This was called the Bose-Einstein condensate, after Einstein and Indian physicist Satyandranath Bose. Wolfgang Ketterle, Eric Cornell, and Carl Wieman were awarded the Nobel Prize in physics in 2001 for creating a Bose-Einstein condensate in their laboratory.

THE SKELETON BONES of Mesozoic dinosaurs (70 million to 220 million years old) were discovered in the Gobi Desert in Mongolia. Sigmund Freud began to publish his *Collected Writings,* which would include twelve volumes by 1939. English astronomer Arthur Eddington, whose experiment had confirmed Einstein's general theory of relativity, discovered that the luminosity of a star is related to its mass.

Edwin Hubble, using the 100-inch telescope on Mount Wilson, measured the distance to the nearest spiral galaxies—two million light-years, a distance so great that our galaxy appeared to shrink overnight. Louis de Broglie, after having offered his thoughts on the nature of matter, waves, and quanta the previous year, predicted that all matter, from electrons to balloons, has the properties of both particles and waves.

J. Edgar Hoover becomes director of the FBI.

The Nobel Prize in physics went to Karl M. G. Siegbahn of Sweden for his discoveries and research in the field of x-ray spectroscopy. The prize money in the chemistry division, meanwhile, was allocated to the section's Special Fund.

122. **"The Compton Experiment" (Das Komptonsche Experiment). *Berliner Tageblatt*, April 20, 1924, suppl.**

Einstein discussed Arthur Compton's discovery of 1922 that an x-ray's wavelength is increased when incident radiation is scattered by free electrons, implying that the scattered quanta have less energy than the quanta of the original beam. This effect is now known as the Compton effect and clearly illustrates the particle concept of electromagnetic radiation. Compton would receive the Nobel Prize in physics in 1927 for this discovery.

123. **"On the One-Hundredth Anniversary of Lord Kelvin's Birth" (Zum hundertjährigen Gedenktag von Lord Kelvins Geburt). *Naturwissenschaften* 12 (1924): 601–602.**

The work of Lord Kelvin (W. Thomson) was based in its entirety on the fundamentals of Newtonian mechanics. Einstein recounted its influence on other researchers, such as James Clerk Maxwell. Instead of giving an overview of Thomson's life, he gave a few examples of the scientist's work that he found particularly appealing.

124. **"Quantum Theory of Monatomic Ideal Gases" (Quantentheorie des einatomigen idealen Gases). *Königlich Preussische Akademie der Wissenschaften* (Berlin). Sitzungsberichte (1924): 261–267. Other treatises on the subject were published in 1925 in the same journal, pp. 3–14 and 18–25.**

Einstein envisioned the penetrating analogy between the properties of thermal radiation and of gases of the "degenerate state," extending the Bose method to the monatomic gases and predicting the Bose-Einstein effect.

125. "On the Ether" (Über den Äther). *Schweizerische Naturforschende Gesellschaft. Verhandlungen* 105, part 2 (1924): 85–93.

126. "On the Theory of Radiometric Forces" (Zur Theorie der Radiometerkräfte). *Zeitschrift für Physik* 27 (1924): 1–6.

1925

After having declined an invitation to lecture in South America in 1922, Einstein accepted the offer in 1925 and embarked on a voyage to Argentina, Uruguay, and Brazil. He was feted there by each country's scientific organizations and German communities.

In solidarity with Gandhi and affirming his own pacifist convictions, Einstein, along with Rabindranath Tagore, H. G. Wells, and others, signed a manifesto against compulsory military service.

In the fall, Einstein became a member of the board of governors of the newly founded Hebrew University in Jerusalem. He soon found himself at odds with other members of the board, which was heavily influenced by American Jewry, who felt they had the right to determine the course of the university because of their large financial contributions. Einstein had visualized a research and teaching university that would take into account the needs of Jews emigrating to Palestine, but the Americans envisioned primarily a teaching college that would give prestigious employment opportunities to those they favored.

Though Einstein had begun to depart from the kind of research conducted by most theoretical physicists, the Royal Society of London awarded him the Copley Medal, its highest honor.

In Germany, Paul von Hindenburg was elected president and Hans Luther chancellor. Hitler published volume 1 of *Mein Kampf,* written while he was imprisoned in Bavaria and published by F. Eher in Munich, as he reorganized the Nazi Party.

CALTECH'S ROBERT A. MILLIKAN discovered the presence of cosmic rays in the upper atmosphere. Samuel Goudsmit and George Uhlenbeck assigned angular momentum to electrons and established that they have a quantum-mechanical property similar to spin. In a seminal paper, Werner Heisenberg changed the relationship between physical concepts and mathematical symbols by using matrices to model quantum systems, thereby establishing "matrix mechanics."

German physicist Wolfgang Pauli formulated his "Pauli exclusion principle"—that no two electrons can be in the same energy state. This helped to explain atomic structure statistically. Following on this, Enrico Fermi and Paul Dirac devised a statistical mechanics valid for a family of particles (fermions, which include electrons) subject to the exclusion principle, now known as Fermi-Dirac statistics.

The Nobel Prize in physics was awarded jointly to James Franck and Gustav Hertz of Germany for their discovery of the laws governing the impact of an electron upon an atom. The chemistry prize went to Richard Zsigmondy of Germany and Austria for his demonstration of the heterogeneous nature of colloid solutions and for the techniques that have become fundamental in modern colloid chemistry.

127. "Mission of Our University." *New Palestine* 8 (1925): 294.
As a new member of the governing board of the Hebrew University of Jerusalem, Einstein outlined the goals of the newly established university for which he had helped raise funds.

128. "Unified Field Theory of Gravitation and Electricity" (Einheitliche Feldtheorie von Gravitation und Elektrizität). *Königlich Preussische Akademie der Wissenschaften* (Berlin). Sitzungsberichte (1925): 414–419.

129. "The Electron and Unified Field Theory" (Elektron und einheitliche Feldtheorie). *Physica* 5 (1925): 330–334.

130. "Quantum Theory of Ideal Gases" (Quantentheorie des idealen Gases). *Königlich Preussische Akademie der Wissenschaften (phys.-math. Klasse)* (Berlin). Sitzungsberichte (1925): 18–25.

Extending the Bose method to monatomic ideal gases, Einstein predicted Bose-Einstein condensation and the rediscovery of wave properties of particles of matter.

131. "Eddington's Theory and the Hamiltonian Principle" (Eddington's Theorie und Hamiltonsches Prinzip). Appendix to German edition of A. S. Eddington, *The Theory of Relativity in Mathematical Perspective (Relativitätstheorie in mathematischer Behandlung)*. Berlin, 1925.

1926

From this point on, Einstein's major contributions in physics began to dwindle, although he continued to publish some scientific papers. In middle age, he would contribute to the debate on quantum mechanics, but his search for a unified field theory was hitting a dead end, though he continued to work on it until his death. (See more on this quest in the section for 1951–52, below.) Because of his accomplishments and charismatic personality, his reputation as the greatest living scientist did not diminish. Indeed, in 1926 the Royal Astronomical Society awarded him its Gold Medal and the Academy of Sciences of the USSR made him an honorary member.

In Germany, the Hitlerjugend fascist youth organization was founded, and the country was admitted to the League of Nations. Paul Joseph Goebbels became the Nazi Party leader in Berlin.

ERWIN SCHRÖDINGER initiated the development of the final quantum theory by describing wave mechanics in terms of a differential equation. He showed complete mathematical equivalence to matrix mechanics. Later in 1926, Max Born stated that the so-called wave function does not describe the behavior of any particle but interpreted the equation in terms of a probability: "quantum mechanics" was completed. Still later in 1926, Werner Heisenberg used unified quantum mechanics to quickly calculate the spectrum of several states of the helium atom.

Robert Goddard finished building a spindly, ten-foot-long liquid-fuel rocket he nicknamed Nell. He set up the missile on his aunt's farm field, and an assistant lit the fuse with a blowtorch attached to a long stick. After some hesitation, the rocket zoomed into the sky at sixty miles per hour to an altitude of forty-one feet and then plummeted into a frozen cabbage patch after two and a half seconds.

Robert Watson-Watt proposed the word *ionosphere* for the upper layer of the atmosphere, where the ionization of gases affects the propagation of radio waves. Gregory Breit and Merle Tuve measured the distance to the ionosphere by measuring the time needed for a radio signal to bounce back.

The Nobel Prize in physics went to Jean Perrin of France for his work on the discontinuous structure of matter, especially his exploration of the sedimentation equilibrium. His experiments confirmed Einstein's 1905 paper on Brownian motion. The chemistry prize was awarded to The Svedberg of Sweden for his work on disperse systems.

132. "On the Cause of the Formation of Meanders in the Courses of Rivers" (Über die Ursache der Mäanderbildung der Flussläufe und des sogenannten Baerschen Gesetzes). *Naturwissenschaften* 14 (1926): 223–334. English translation in *The World as I See It* (1934; not in the abridged edition) and *Ideas and Opinions* (1954).

Read first before the Prussian Academy of Sciences on January 7, this paper discussed why the erosion of rivers tends to occur on the right bank in the Northern Hemisphere and on the left in the Southern Hemisphere. Having found no one who was thoroughly familiar with the causal relations involved, Einstein decided to tackle the problem himself.

133. "Suggestion for an Experiment on the Nature of the Elementary Radiation Emission Process" (Vorschlag zu einem die Natur des elementaren Strahlungs-Emissionsprozesses betreffendes Experiment). *Naturwissenschaften* 14 (1926): 300–301.

134. "Interference Characteristics of Light Emitted by Canal Rays" (Über die Interferenzeigenschaften des durch Kanalstrahlen emittierten Lichtes). *Königlich Preussische Akademie der Wissenschaften* (Berlin). Sitzungsberichte (1926): 334–340.

1927

The First Solvay Congress in Brussels in 1911 was a who's who of physics. Einstein met with Marcel Brillouin, Marie Curie, Paul Langevin, Hendrik Lorentz, Jean Perrin, Max Planck, and Ernest Rutherford.

In the fall, at the latest Solvay Congress in Brussels, Einstein and Niels Bohr began to spar about the foundations of quantum mechanics. The debate would continue until Einstein's death.

Despite Einstein's opposition, his son Hans Albert, at age twenty-three, married Frieda Knecht, a woman nine years his senior. Even Mileva felt that Frieda was not good enough, writing a friend the following year that Hans Albert looked terrible and that "his wife does not know how to look after him, she thinks only of herself." Frieda, a warm and bright woman, bore three children—David, Klaus, and Bernhard—only one of whom (Bernhard) lived beyond the age of six. She and Hans Albert adopted a baby girl, Evelyn, in 1941.

In November, Einstein, along with inventor and physicist Leo Szilard, applied for a patent on a home refrigerator using what has come to be known as the Einstein-Szilard pump. The refrigerators of the time were noisy and unreliable; worse, they often poisoned their owners by leaking toxic refrigerants. Proposing a nonmechanical absorption refrigerator, Einstein and Szilard sold this idea and a later one to a division of the Electrolux Company. AEG, the German division of General Electric, bought a third design. In all, the two collaborated on five designs. But by 1932 the development of these refrigerators was abandoned because of the economic downturn and the invention of Freon, a safer coolant. The two scientists continued their seven-year collaboration to make and patent other mechanical gadgets. Pumps such as the Einstein-Szilard model are now used to circulate liquid sodium coolant in nuclear reactors.

Szilard, a native of Hungary, was noted for his contributions to nuclear physics, thermodynamics, atomic energy, and molecular biology. His ideas and patents included the cyclotron, linear accelerator, electron microscope, and nuclear chain reaction. The adverse economic and political times during which he worked prevented the development and manufacturing of his inventions. Others later expanded on his ideas and received credit for them—even Nobel Prizes. Szilard was instrumental in drafting the famous letter from Einstein to President Franklin D. Roosevelt in August 1939, recommending an atomic energy program, which led to the Manhattan Project and the eventual building of the atomic bomb in 1945. Szilard circulated a petition that year asking that the United States not use the bomb against Japan on moral grounds.

In Germany, the stock market crashed in May on Black Friday and the economic system collapsed.

PAUL DIRAC DESCRIBED a method of quantizing the electromagnetic field, and Niels Bohr and Werner Heisenberg, a junior associate, discussed what came to be known as the Copenhagen interpretation of quantum mechanics; Bohr took the position that the measurement process of a specific experimental situation engenders the impossibility of simultaneously measuring a particle's position and its momentum. Einstein opposed the Copenhagen school of thought.

George Paget Thomson diffracted electrons by passing them through a thin foil in a vacuum, thereby verifying de Broglie's wave hypothesis. Donald Menzel obtained the first accurate measurements of the surface temperatures of Mars and Mercury.

Charles Lindbergh flew *The Spirit of St. Louis* nonstop from New York to Paris in thirty-three and a half hours.

The Nobel Prize in physics was shared by Arthur Compton of the United States, for the effect named after him, and Charles Wilson of Britain, for his cloud chamber, which makes the paths of electrically charged particles visible by the condensation of vapor. The chemistry prize was awarded to Heinrich O. Wieland of Germany for his investigations of the constitution of bile acids and related substances.

---

135. "On Kaluza's Theory on the Connection between Gravitation and Electricity" (Zur Kaluzas Theorie des Zusammenhanges von Gravitation und Elektrizität). *Königlich Preussische Akademie der Wissenschaften* (Berlin). Sitzungsberichte (1927): 23–35.

Kaluza attempted to unify Einstein's theory of gravity with Maxwell's theory of light by mixing them in a fifth dimension. He worked with Oskar Klein, a mathematician, and they produced the Kaluza-Klein field equations.

---

136. "Effect of Earth's Motion on the Velocity of Light Relative to Earth" (Einfluss der Erdbewegung auf die Lichtgeschwindigket relativ zur Erde). *Forschungen und Fortschritte* 3 (1927): 36–37.

---

137. "Formal Relationship of the Riemann Curvature Tensor to the Field Equations of Gravitation" (Über die formale Beziehung des Riemannschen Krümmungstensors zu den Feldgleichungen der Gravitation). *Mathematische Annalen* 97 (1927): 99–103.

---

138. "Newton's Mechanics and Its Influence on the Shaping of Theoretical Physics" (Newtons Mechanik und ihr Einfluss auf die Gestaltung der theoretischen Physik). *Naturwissenschaften* 15 (1927): 273–276. For English translation, see *The World as I See It* (1934; not in the abridged edition) or *Ideas and Opinions* (1954).

Writing on the occasion of the two-hundredth anniversary of Newton's death, Einstein traced the historical development of physics from the Greeks to Galileo to Newton to the present, especially emphasizing Newton's contributions.

139. "Isaac Newton." Letter to the Royal Society on the two-hundredth anniversary of Newton's death. *Nature* 119 (1927): 467.

In this letter, Einstein praised the British for their traditions and for providing an atmosphere that allows the human soul "to soar." Everything that had happened in theoretical physics since Newton's time had developed from his idea, and only in quantum theory was Newton's differential method inadequate.

140. "General Theory of Relativity and the Laws of Motion" (Allgemeine Relativitätstheorie und Bewegungsgesetze). *Königlich Preussische Akademie der Wissenschaften* (Berlin). Sitzungsberichte (1927): 235–245.

Einstein attacked the problem of deducing the law of motion from the field equations. He repeatedly returned to this problem in his later work.

1928

In March, while on a trip to Davos, Switzerland, the forty-nine-year-old Einstein collapsed with a serious heart condition. He was confined to bed for four months, but it took him a year to recover fully.

In April, Einstein hired Helen Dukas as his last secretary. She would remain with him, as part of his household, for the remainder of his life, and after his death she became the archivist of his collected papers at the Institute for Advanced Study in Princeton. Fiercely loyal to him, she was largely responsible for protecting his image as a lovable, benevolent, absent-minded genius and for keeping his personal life private.

Einstein was elected to the board of directors of the German League for Human Rights, a pacifist organization.

The two papers listed below were submitted to the Prussian Academy of Sciences by Max Planck, since Einstein was too ill to attend the meetings. In these papers he presented a new mathematics that he felt would allow him to formulate his unified theory of gravity and electricity.

THIS WAS A PRODUCTIVE YEAR in physics. George Gamow devised the "liquid drop model" of the atomic nucleus, implying that something like surface tension holds the nucleus together. Leo Szilard and, independently, Rolf Wideröe proposed advanced designs for a linear accelerator. Hermann Weyl created a matrix theory of continuous groups and discovered that many of the regularities of quantum phenomena could best be understood by means of group theory. German physicist Hans Geiger and graduate student Walter Müller invented the first Geiger counter to detect radioactive particles. Chandrasekhar Raman predicted how light scattered from a material can be used to gather information about the properties of that material, which now forms the basis of a powerful spectroscopic tool. Finally, mathematician John von Neumann conceived game theory.

The Nobel Prize in physics was awarded to England's Sir Owen W. Richardson for his work on thermionic phenomena and especially for the discovery of the law named after him. Germany's Adolf Windaus received the chemistry prize for his research into the constitution of sterols and their relation to vitamins.

141. "Riemannian Geometry with Preservation of Distant Parallelism" (Riemann-Geometrie mit Aufrechterhaltung des Fernparallelismus). *Königlich Preussische Akademie der Wissenschaften* (Berlin). Sitzungsberichte (1928): 217–221.

142. "New Possibilities of a Unified Field Theory of Gravitation and Electricity" (Neue Möglichkeiten für eine einheitliche Feldtheorie von Gravitation und Elektrizität). *Königlich Preussische Akademie der Wissenschaften* (Berlin). Sitzungsberichte (1928): 224–227.

143. "H. A. Lorentz." *Mathematisch-naturwissenschaftliche Blätter* 22 (1928): 24–25. Reprinted as "Address at the Grave of H. A. Lorentz" in *Ideas and Opinions* (1954), 73.

In this eulogy at the graveside of the Dutch theoretical physicist whom Einstein admired and loved, Einstein called him the "greatest and noblest man of our time."

1929

On March 14, Einstein celebrated his fiftieth birthday and was showered with congratulatory cards, letters, and telegrams sent by young and old from all over the world. In honor of this occasion, his friends were anxious to obtain a substantial gift for him: a summer home to call his own, to be donated by the city of Berlin. The plan was foiled by politics and property rights, however, and, tired of witnessing one fiasco after another, Einstein decided to buy his own plot of land and build a summer home in Caputh, three minutes from a lake where he could sail. His friends then surprised him with the gift of a sailboat. The house was built quickly, and the Einsteins occupied it in September, though they kept the apartment in Berlin as well. Einstein exulted in the rural setting, and his health improved quickly. With renewed energy, he carried on an extensive series of interviews and correspondence on behalf of Zionist and pacifist goals. Most importantly, however, he published a paper on the unified field theory that was a sensation even before publication (see paper 145 below).

In June Einstein received the Planck Medal from the German Physical Society.

He began a friendship and correspondence with Queen Elisabeth of Belgium and her husband, King Albert. During a visit to his uncle Caesar, Einstein had been invited to meet the queen in her palace at Laeken, where they played music as a trio with a lady-in-waiting. The queen charmed Einstein with her simplicity and cordiality, and he visited her and her husband again a year later. Theirs became a warm and enduring friendship, and Einstein corresponded with the queen until the end of his life; the king died in 1934. Einstein always began his letters to her with "Dear Queen."

In Germany, Hitler appointed Heinrich Himmler as *Reichsführer* of the SS. The World Zionist Organization established the Jewish Agency to represent all Jews, both Zionist and non-Zionist, to fulfill the League of Nations'

Children living in a hostel for the sons and daughters of the working poor send Einstein wishes for a happy fiftieth birthday.

Einstein and Marie Curie

mandate that a "Jewish agency" comprising representatives of world Jewry assist in the "establishment of the Jewish Home ... in Palestine."

PAUL DIRAC PUBLISHED his "relativistic wave equation," which describes the spin of electrons and led to the prediction of the electron's antiparticle, the positron. Edwin Hubble published a seminal paper outlining his observations that galaxies are receding, which caused Einstein to abandon the cosmological constant as the "biggest blunder of my life" (see discussion of paper 67, above). And Sir Frank Whittle, combining the concepts of rocket propulsion and gas turbines, invented jet propulsion.

The romantically named Louis-Victor Pierre Raymond de Broglie, a French prince, received the Nobel Prize in physics for his work showing that matter, specifically the electron, can sometimes exhibit wavelike properties. The chemistry prize was divided equally between another romantically named scientist, Hans Karl August Simon von Euler-Chelpin of Sweden and Germany, and Sir Arthur Harden of England for their investigations on the fermentation of sugars and enzymes.

144. [Text of broadcast on the semicentennial of Thomas A. Edison's incandescent lighting.] Printed in the *New York Times*, October 23, 1929.

145. "On the Unified Field Theory" (Zur einheitlichen Feldtheorie). *Königlich Preussische Akademie der Wissenschaften* (Berlin). Sitzungsberichte (1929): 2–7.

This exposition of Einstein's theory on unitary field laws for gravitation and electromagnetism centered on his efforts to incorporate the electromagnetic field into the geometry of space-time. The publication of this paper was immediate news. Gossip in the mass media, based on the two papers he published in 1928, had claimed that Einstein had solved the riddle of the universe, and this publication was highly awaited. The first printing immediately sold out, and more printings were ordered. Einstein could not understand the commotion, even though he had faith in his theory. Critics found fault with it, however, and after a few more attempts at corrections, Einstein finally admitted, three years later, that he was wrong.

146. "The New Field Theory." *New York Times*, February 3, 1929.
A report on paper 145.

147. "Unified Interpretation of Gravitation and Electricity" (Einheitliche Interpretation von Gravitation und Elektrizität). *Königlich Preussische Akademie der Wissenschaften* (Berlin). Sitzungsberichte (1929): 102.

148. "Unified Field Theory and the Hamiltonian Principle" (Einheitliche Feldtheorie und Hamiltonsches Prinzip). *Königlich Preussische Akademie der Wissenschaften* (Berlin). Sitzungsberichte (1929): 156–159.

1930

For the first time, Einstein became a grandfather when Hans Albert's wife, Frieda, gave birth to a son, Bernhard; he became a father-in-law for the third time when Margot tied the knot with Dmitri Marianoff (the marriage later ended in divorce). His other son-in-law, Rudolf Kayser, published the first biography of Einstein; Marianoff would publish another fourteen years later. Albert and Elsa prepared for another trip to America, this time to California, where Einstein would lecture at the California Institute of Technology in Pasadena.

As the Nazis gained the majority in the German elections and Heinrich Brüning formed a right-wing coalition government, Einstein stepped up his political ac-

Einstein's theory of relativity is alleged to be simpler than the tax code.

tivities in the face of increasing German nationalism and unrest. As a socialist, his politics were left of center. Although he was not a Communist, he was not averse to contact with them, being curious about all points of view. He drew the line at participating in meetings at which Communists dominated, however.

Einstein attended his last Solvay Congress in Brussels and received an honorary degree from the ETH (Poly) in Zurich.

In December, their ship was anchored in New York harbor for five days before continuing its voyage, and the Einsteins again faced the press, met celebrities, and returned exhausted at night to their stateroom. Among others, they met Arturo Toscanini, Fritz Kreisler, and John D. Rockefeller Jr. Einstein visited the Riverside Church on Manhattan's Upper West Side to view his image and those of many others sculpted into the tympanum over the new church's main entrance—a sight that delighted him, according to the church's rector.

As the ship sailed south and then west toward California, the Einsteins stopped in Cuba for two days before passing through the Panama Canal.

Both in New York and in California, Einstein spoke out in favor of pacifism and antimilitarism, usually to the embarrassment of his hosts.

In the Soviet Union and its satellites, anti-Semitism was growing. Palestine, which had been ruled by the British since World War I, was beginning to receive huge waves of eastern European Jews—about 200,000—coming from the *shtetls* to resettle the land of their ancestors. The British were aware of Arab resistance to such a huge influx and recommended the cessation of Jewish immigration. At this time, about 9.5 million Jews lived in Europe, mostly in the cities. Those in Germany, about 0.5 million in a German population of about 65 million, were the most assimilated in Europe; for the most part they had no desire to join the early exodus. In Germany, the Nazi Party had emerged as the majority party in the national elections, portending as yet unknown evils.

ARTHUR EDDINGTON attempted to unify the general theory of relativity with quantum theory. American physicist Ernest O. Lawrence published the principle of the cyclotron, permitting the acceleration of atoms to high speeds to create nuclear reactions. Subrahmanyan Chandrasekhar of India, while a student at Cambridge, calculated around this time that stars with a core mass of less than 1.4 Suns will eventually collapse under their own weight, forming a dense state known as a white dwarf. This finding, called the "Chandrasekhar limit," paved the way for the theoretical prediction of neutron stars and black holes for stars whose mass is above this limit.

The Nobel Prize in physics was awarded to Sir Chandrasekhar Venkata Raman of India (Chandrasekhar's uncle) for his work on the scattering of light and the discovery of the effect named after him. Hans Fischer of Germany received the chemistry prize for his research into the constitution of hemin and plant pigments (chlorophyll), especially his synthesis of hemin.

149. "The Compatibility of Field Equations in the Unified Field Theory" (Die Kompatibilität der Feldgleichungen in der einheitlichen Feldtheorie). *Königlich Preussische Akademie der Wissenschaften* (Berlin). Sitzungsberichte (1930): 18–23.

150. "Two Exact Statistical Solutions of the Field Equations of the Unified Field Theory" (Zwei strenge statistische Lösungen der Feldgleichungen in der einheitlichen Feldtheorie) (with Walther Mayer). *Königlich Preussische Akademie der Wissenschaften* (Berlin). Sitzungsberichte (1930): 110–120.

151. "On Progress Made by the Unified Field Theory" (Über die Fortschritte der einheitlichen Feldtheorie) (a summary). *Königlich Preussische Akademie der Wissenschaften* (Berlin). Sitzungsberichte (1930): 102.

152. "On the Theory of Spaces with the Riemann Metrics and Distant Parallelism" (Zur Theorie der Räume mit Riemann-Metrik und Fernparallelismus). *Königlich Preussische Akademie der Wissenschaften* (Berlin). Sitzungsberichte (1930): 401–402.

153. "The Unified Field Theory Based on Riemann Metrics and Distant Parallelism" (Auf die Riemann-Metrik und den Fernparallelismus gegründete einheitliche Feldtheorie). *Mathematische Annalen* 102 (1930): 685–697.

154. "Space, Ether, and Field in Physics" (Raum, Äther und Feld in der Physik). *Forum Philosophicum* 1 (1930): 173–180.

155. "About Kepler" (Über Kepler). *Frankfurter Zeitung*, November 9, 1930. English translation in *The World as I See It* (1934) and *Ideas and Opinions* (1954).

Writing on the occasion of the three-hundredth anniversary of Johannes Kepler's death, Einstein recounted the difficulties Kepler, working on planetary motion, must have faced in his time and praised his ingenious methods. He concluded that "knowledge cannot spring from experience alone but from the comparison of the inventions of the intellect with observed fact."

156. "Religion and Science." *New York Times*, November 9, 1930. Reprinted in *Ideas and Opinions* (1954), 36–40. See also papers 216, 226, and 257 on the same topic.

Einstein wrote this article expressly for the *New York Times Magazine*. It was reprinted as the essay "Cosmic Religion" in the book of the same name the following year (see no. 166 below). In this somewhat provocative piece, Einstein professed his belief in a "cosmic religion," which to him was a higher level of religion than organized religion. He declared it to be a "miraculous order that manifests itself in all of nature as well as in the world of ideas," devoid of a personal God who doles out rewards and punishments based on people's behavior. He concluded that there is no conflict between science and religion, indeed, that cosmic religiosity is necessary for scientific research.

157. "Science and God: A Dialogue." *Forum and Century* 83 (1930): 373–379.

This article is based on a conversation with J. Murphy and J. W. N. Sullivan, touching on the relationship of science to other aspects of life, the question of Jewish racial characteristics, and other topics of interest to Einstein. Sullivan was a mathematician and science popularizer who had undertaken a "tour of great men," interviewing as many eminent scientists as he could find for an upcoming book. J. Murphy may have been Joseph Murphy, a world-renowned authority on mysticism and metaphysics, or James Murphy (see paper 181 below).

158. "What I Believe." *Forum and Century* 84 (1930): 193–194. Reprinted as "The World as I See It" in *Ideas and Opinions* (1954), 8–11.

Here Einstein put forth his personal beliefs and philosophy. This piece contains his famous quotations: "I have never looked upon ease and happiness as ends in themselves—such an ethical basis I call the ideal of a pigsty . . . The ideals which have guided my way . . . have been Kindness, Beauty, and Truth," and "The most beautiful experience we can have is the mysterious."

159. "Concept of Space." *Nature* 125 (1930): 897–898.

160. "On the Present State of the Theory of Relativity." *Yale University Library Gazette* 6 (1930): 3–6.

161. "The Space-Time Problem" (Das Raum-Zeit Problem). *Koralle* 5 (1930): 486–488.

162. "Speech at the Broadcasting Exhibition" (Rede zur Funkausstellung). Berlin, August 22, 1930. Transcribed from the sound recording by F. Herneck. *Naturwissenschaften* 49 (1930): 33.

1931

In January, Einstein was the guest of Charlie Chaplin at a Hollywood premiere of the film *City Lights*. The picture of the two charismatic men together was circulated widely in the media. Einstein met other celebrities as well, including Helen Keller, Upton Sinclair, and Norman Thomas. He also went to Mount Wilson Observatory to meet with astronomer Edwin Hubble, who in 1929 had obtained proof that the universe was indeed expanding, causing Einstein to dismiss the cosmological constant of his 1917 paper as his "biggest blunder." (But see discussion above under paper 67.)

The Empire State Building nears completion.

The return trip from California to New York in March was by train, with a stop at a Hopi Indian reservation near the Grand Canyon. The Hopis gave Einstein a peace pipe, recognizing his pacifism, and dubbed him their "Great Relative." During a short stop in Chicago, Einstein had just enough time to fire off yet another pacifist speech. At New York's Astor Hotel, he gave a speech at a fund-raiser for Palestine, urging Jews to cooperate with Arabs. Then it was back on board ship for the return to Germany, which was now on shaky grounds politically.

After having barely recuperated from his trip and delivering two papers to the Prussian Academy of Sciences, Einstein took leave again in May for a month's stay in

Albert Einstein, "The Great Relative," visits with Hopi Indians at the Grand Canyon.

England, where he delivered the Rhodes Lectures at Oxford and accepted an honorary degree. The blackboard on which Einstein wrote some equations at Oxford has been preserved. Einstein's equations from a lecture on June 6, 1930, in Nottingham have also been preserved at its university.

During the summer, he sequestered himself at the house in Caputh. Ever concerned about war, he wrote letters and issued statements on pacifism and the need to refuse military service. He was convinced that Germany was heading toward an aggressive dictatorship under Adolf Hitler.

In December, Einstein was again en route to Pasadena, on a ship that took him and Elsa directly to Los Angeles. He had decided to leave Germany and was negotiating employment elsewhere, including at Caltech, though he would have preferred staying in Europe. He continued to speak out in favor of pacifism and commented on racial discrimination in the United States.

As the Great Depression became a worldwide phenomenon, German banks closed down.

Pacifist Mahatma Gandhi travels to London to make his case for Indian independence.

IN THE SCIENCES, Wolfgang Pauli, while in Zurich, predicted the existence of "a little neutral thing," the neutrino, to explain where the energy went during beta decay. Bernhard V. Schmidt invented a new type of telescopic optical system enabling astronomers to take sharp photographs of large areas of the sky. Austrian logician Kurt Gödel published, at age twenty-five, his famous "incompleteness theorem," showing that within any given branch of mathematics there would always be some propositions that cannot be proven either true or false using the rules and axioms of that branch. Ernest Lawrence's cyclotron, which would later be called an "atom smasher," was finished in Berkeley.

The Nobel Prize money for the physics section was allocated to the Main Fund of the foundation and to the Special Fund in physics. The prize in chemistry was given jointly to Germany's Friedrich Bergius and Carl Bosch for the development of chemical high-pressure processes.

163. "Science and Dictatorship." In Otto Forst-Battaglia, ed., *Dictatorship on Trial*, translated by Huntley Paterson, introduction by Winston Churchill. London: George G. Harrop, 1930. Reprinted in 1970 by Books for Libraries Press, Freeport, N.Y.

In this book, compiled at the time of Hitler, Stalin, and Mussolini, "eminent leaders of modern thought" contributed their ideas on the expedient of dictatorship throughout history. Einstein's contribution consists of only two lines, in which he says that dictatorship means that people are muzzled, and therefore stultified, and that "science can flourish only in an atmosphere of free speech."

164. "Science and Happiness." Speech at California Institute of Technology. Reprinted in *New York Times*, February 22, 1931, sec. 9. Also published in *Science*, new series, 73, no. 1893 (April 10, 1931): 375–381.

Einstein rhetorically asked why science has brought so little happiness. His answer: We have not yet learned to make sensible use of it. Scientists must not forget that concern for people's happiness is paramount in all technical endeavors.

165. "Militant Pacifism." *World Tomorrow* 14 (1931): 9. Reprinted in *Ideas and Opinions* (1954) as "Active Pacifism."

Einstein made these remarks in response to a peace demonstration in Flemish Belgium, calling for disarmament and hoping that future generations "will look back on war as an incomprehensible aberration of their forefathers."

166. *Cosmic Religion, with Other Opinions and Aphorisms*. New York: Covici-Friede, 1931.

This small book contains a biographical note, essays on religion, pacifism, and the Jews, and aphorisms taken from earlier sources.

167. "Tagore Talks with Einstein." *Asia* 31 (1931): 138–142.

This conversation on Eastern music was reported by Indian poet and musician Rabindranath Tagore.

168. "The Nature of Reality." *Modern Review* (Calcutta) 49 (1931): 42–43.

In this conversation, Einstein and Tagore discussed the nature of beauty and truth.

169. "The 1932 Disarmament Conference." *Nation* 133 (1931): 300. Reprinted in *Ideas and Opinions* (1954).

After declaring that the state should be the servant of man and not vice versa, Einstein discussed the need for international disarmament, abolishment of compulsory military service, and changes in educational systems that pass on military traditions. He decried nationalism as unhealthy, leading to aggression and war, and called for the protection of conscientious objectors worldwide. In an optimistic tone, he wrote that he believed that responsible national leaders "do, in the main, honestly desire to abolish war."

170. *About Zionism: Speeches and Letters*. Translated and edited by Sir Leon Simon. New York: Macmillan, 1931.

Einstein's speeches and letters about Zionism were collected from various sources and dated. Unfortunately, precise locations in the original sources are not given.

171. "On the Cosmological Problem of the General Theory of Relativity" (Zum kosmologischen Problem der allgemeinen Relativitätstheorie). *Königlich Preussische Akademie der Wissenschaften* (Berlin). Sitzungsberichte (1931): 235–237.

In this paper Einstein accepted the nonstatic character of the universe and rejected the cosmological constant as unnecessary and as compromising the simplicity of his field equations. (Also see the discussion under paper 67 above.)

172. "Unified Theory of Gravitation and Electricity" (Einheitliche Theorie von Gravitation und Elektrizität), part 1 (with Walther Mayer). *Königlich Preussische Akademie der Wissenschaften* (Berlin). Sitzungsberichte (1931): 541–557 (part 2 published in 1932).

173. "Foreword" to Sir Isaac Newton, *Opticks*. New York: McGraw-Hill, 1931.

Reviewing Newton's work, Einstein summarized that "in one person he combined the experimenter, the theorist, the mechanic, and, not the least, the artist of exposition."

174. *Relativity: The Special and General Theory.* Translated by R. Lawson. New York: Peter Smith, 1931.

According to Einstein, the book was intended "to give an exact insight into the theory of relativity to those readers who, from a general and philosophical point of view, are interested in the theory, but who are not conversant with the mathematical apparatus of theoretical physics." Rather than writing elegantly, he repeated himself frequently "in the interest of clearness."

1932

While at the California Institute of Technology from January until March, Einstein met Abraham Flexner, a prominent educational reformer, fund-raiser, and organizer who was interested in innovative and progressive educational institutions. (See Thomas Bonner's biography of Flexner, which is listed in the bibliography.) Flexner was the founding director of the Institute for Advanced Study in Princeton, New Jersey, which was established in 1930 through an endowment by Louis Bamberger and his sister, Caroline Fuld. Flexner was looking for a faculty for the new institute and was eager to get Einstein on board. He offered an attractive position—a half-year, annual appointment, which would enable Einstein to spend the other half-year in Berlin, as he was still not quite ready to give up his house in Caputh and, per-

Amelia Earhart completes her transcontinental flight.

haps, his new love, Toni Mendel. He agreed to come to Princeton in October of the following year.

Einstein thereafter traveled to Geneva for a peace conference, visited Oxford, and corresponded with Sigmund Freud and Maxim Gorki.

In the meantime, he became a grandfather for the second time when Frieda and Hans Albert's baby, Klaus, was born in Zurich. Little Klaus died six years later.

In Germany, Paul von Hindenburg was elected president, defeating Hitler by 7 million votes, but the Nazis took control of the parliament (the Reichstag). Franz von Papen was named chancellor when Hitler refused Hindenburg's offer to become vice chancellor; then van Papen resigned and Kurt von Schleicher was named to the post. Hitler, who was born in Austria, received German citizenship.

As a great famine swept the USSR, Jews, mostly Zionists, continued their exodus to Palestine to establish agrarian settlements and begin a new life in their promised land.

IT WAS A BUSY YEAR in physics. Early in 1932 Marie's daughter Irène Curie and Frédéric Joliot bombarded nonradioactive beryllium with alpha particles, briefly turning it into a radioactive element. Later that year, Enrico Fermi of Italy succeeded in intensifying the Curie-Joliot effect by using the newly discovered and massive neutrons of beta rays instead of alpha rays. Meanwhile, Werner Heisenberg proposed a new model of the atom in which protons and neutrons achieve stability by exchanging electrons. James Chadwick isolated the neutron, the first particle discovered with zero electrical charge, though a two-year debate ensued as to whether the new particle was really a fundamental building block or a composite of the proton and electron.

This year also saw the announcement of the first apparatus that could artificially accelerate atomic particles to high energies—the Cockcroft-Walton accelerator. Within a month, beams of high-energy protons pro-

Einstein with Robert Millikan (left) and Georges Lemaître, three giants of modern physics.

duced by the machine were used to initiate the disintegration of lithium nuclei, confirming the equivalence of mass and energy, as Einstein had predicted.

Werner Heisenberg of Germany won the Nobel Prize in physics for the creation of quantum mechanics, whose application led to the discovery of allotropic forms of hydrogen. America's Irving Langmuir received the chemistry prize for his discoveries and investigations in surface chemistry.

175. "To American Negroes." *Crisis* 39 (1932): 45.

In this article written for the NAACP's official journal, Einstein criticized racism after he had witnessed rampant racial prejudice during his visit to the United States in the winter of 1931–32.

176. "Is There a Jewish View of Life?" *Opinion* 2 (September 26, 1932): 7. Reprinted in *Ideas and Opinions* (1954).

In answer to this question, Einstein replied that in his opinion there is no specific Jewish point of view. But he believed the Jewish tradition has a reverence for life and a positive attitude toward it and contains "a sort of intoxicated joy and amazement at the beauty and grandeur of this world."

177. "Unified Theory of Gravitation and Electricity" (Einheitliche Theorie von Gravitation und Elektrizität), part 2 (with Walther Mayer). *Königlich Preussische Akademie der Wissenschaften* (Berlin). Sitzungsberichte (1932): 130–137.

Walther Mayer was the young assistant who later immigrated to the United States with Einstein. He had joined Einstein in 1930 and, of all of Einstein's collaborators, he coauthored the largest number of papers with him. Mayer continued to work with Einstein for a short time at the Institute for Advanced Study in Princeton.

178. "Semivectors and Spinors" (Semi-Vektoren und Spinoren) (with Walther Mayer). *Königlich Preussische Akademie der Wissenschaften* (Berlin). Sitzungsberichte (1932): 522–550.

179. "On the Relation between the Expansion and the Mean Density of the Universe" (with Willem de Sitter). *Proceedings of the National Academy of Sciences* (USA) 18 (1932): 213–214.

Einstein and de Sitter put forth a revised cosmological model that solves the Friedmann equations and takes account of Edwin Hubble's evidence for the expansion of the universe, tentatively implying an initial singularity.

180. "Present State of Relativity Theory" (Gegenwärtiger Stand der Relativitätstheorie). *Die Quelle* (later *Pädagogischer Führer*) 82 (1932): 440–442.

181. "Prologue and Epilogue: A Socratic Dialogue" (with James Murphy). In Max Planck, *Where Is Science Going?* New York: Norton, 1932.

182. "Introduction and Address to Students of UCLA," February 1932. In *Builders of the Universe*. Los Angeles: U.S. Library Association, 1932.

In this little-known text presented in both English and German, Einstein discussed the source of his own creative scientific work. The subject is science as a coordination of observed facts, as seen, for example, in the progression from the special theory of relativity to the unified field theory.

183. "On Dr. Berliner's Seventieth Birthday" (Zu Dr. Berliners siebzigsten Geburtstag). *Naturwissenschaften* 20 (1932): 913. Reprinted in *Ideas and Opinions* (1954), 68–70.

Berliner, a German physicist and Jew, was editor of the German physics journal *Naturwissenschaften* from 1913 to 1935, when the Nazis dismissed him. In 1942, when he was about to be deported, he committed suicide at the age of eighty. In this article, Einstein praised Berliner's work on the journal.

1933

Hitler Youth salute their approving Führer.

While Einstein and Elsa were still in Pasadena, their lives were drastically affected by world events. In January, the Nazis came to power, with Hitler at the helm as chancellor, and the purging of Jews began. Einstein was advised that it was now too dangerous to return to Germany, since a price had been put on his head and his books were being discredited and burned. The Einsteins went to Belgium and set up a temporary residence at Coq sur Mer, a peaceful seaside village surrounded by dunes. Security guards were sent by the Belgian government to protect them. In April, Helen Dukas and Walther Mayer arrived from Berlin to join them. Einstein's stepdaughter Margot and her husband Dmitri had already fled to Paris, and Ilse and Rudolf were still in Berlin. During this stay in Belgium, Einstein became better acquainted with Queen Elisabeth.

Einstein's house and bank accounts were confiscated by the Nazis, but he had kept funds in foreign banks in anticipation of this crisis. He became disillusioned with pacifism and warned the world repeatedly that Germany was preparing for war. He reluctantly decided that countries that are in grave danger, such as Belgium, had no choice but to depend on their armed forces in the face of a formidable and vicious enemy. Under current conditions, he maintained, he would gladly serve in the military to defend European civilization. At heart he was still a pacifist, he said, but that was a position that one could take only when military dictatorships ceased to exist. In response to his change of opinion, pacifist leaders accused him of being a turncoat.

Einstein resigned from all of his positions and affiliations in Germany and gave up his German citizenship. After considering several job offers, he decided to accept the one proposed by Flexner, which had in the meantime been upgraded to a regular, year-round position in Princeton.

The itinerant Einstein spent the month of June in Oxford again, delivering a lecture during his stay, and this time he met Winston Churchill. Then he made a final trip to Switzerland and saw his son Eduard for the last time. Eduard had developed schizophrenia and was confined to a mental institution at Burghölzli, where he stayed periodically until he died there in 1965. Einstein, feeling that he could do nothing to help his son other than offer him financial security, returned to England in September for four weeks before embarking for the United States. He gave one last speech in Europe, at the Royal Albert Hall in London on October 3, warning his listeners about the dangers that lay ahead and praising them for remaining loyal to their democratic traditions.

Ilse's husband, Rudolf Kayser, managed to have Einstein's writings and correspondence and some furnishings, including the grand piano, sent from Berlin to France. From there, everything eventually made its way to the United States by way of diplomatic channels.

The small Einstein entourage, without Ilse and Margot and their spouses, who stayed in Paris, arrived in Princeton on October 17. They spent their first night at the quaint Peacock Inn before moving into temporary quarters on the corner of Library and Mercer streets. The residence was about halfway between the university's mathematics building, where Einstein would have a temporary office, and the grounds of the Institute for Advanced Study, which had yet to be built. A permanent home on Mercer Street not far from the temporary home would become available a year and a half later.

Einstein now began his "American years" as the first faculty member of the Institute for Advanced Study. He never returned to Europe.

As politics slipped into the extremes, with the fascist Nazis on one side and the hardline Communists on the other, the Nazis ordered the building of the first concentration camps in Germany at Dachau near Munich to incarcerate Communists. By 1945, 8 to 10 million Jews and other "undesirables" had been interned and at least half were killed. Germany was no longer interested in peace and quit the League of Nations, withdrawing from all disarmament talks. Hitler was granted dictatorial powers of the Third Reich and suppressed all other political parties. He ordered the burning of all books written by Jews and non-Nazis, including those of Einstein and Freud, demanding total conformity and condemning liberal ideals. Racial hysteria and prejudice drove many Nazis into fanaticism, and any Germans who persisted in espousing liberalism were to be destroyed—chiefly Jews and Communists. All modernist art was suppressed in favor of superficial realism, and scientific research was seriously hampered by new regulations. Before 1939, sixty thousand artists and hundreds of scientists managed to leave Germany. While the Nazis continued to terrorize the Austrians, Austrian chancellor Engelbert Dollfuss opposed union *(Anschluss)* with Germany, and the following year he would pay for it with his life.

In the USSR, famine and starvation reached disastrous proportions and Jews continued to leave for Palestine in search of a better life. Stalin, not to be outdone by Hitler in his inhumanity, began his great purge of the

Paul Dirac (left), Werner Heisenberg, and Erwin Schrödinger

Communist Party, arresting, imprisoning, and executing many Bolsheviks and Trotskyist sympathizers, most of them intellectuals, throughout the USSR.

AUSTRIAN PHYSICIST Erwin Schrödinger and British physicist Paul A. M. Dirac shared the Nobel Prize for their investigations of quantum mechanics. Schrödinger developed an equation to describe nonrelativistic particles. Dirac took into account particles with spin and predicted the existence of antimatter. The prize for chemistry was allocated to the foundation's Main Fund and to the chemistry section.

184. *Why War?* Translated by Stuart Gilbert. Paris: International Institute of Intellectual Cooperation, League of Nations, 1933.

This slight pamphlet constitutes an exchange of letters between Einstein and Sigmund Freud on the proclivity of humankind to make war and what action one can take to counter it.

185. *The Fight against War.* Edited by Alfred Lief. New York: John Day, 1933.

This sixty-four-page pamphlet contains selections from Einstein's writings and speeches on war covering the years 1914–32. The historical setting and usually a specific reference are given for each statement.

186. "On German-American Agreement" (Zur deutsch-amerikanischen Verständigung). *California Institute of Technology, Bulletin* 43, no. 138 (1933): 4–8 (in German), 9–12 (in English).

From a symposium broadcast on January 23, 1933. See also the *New York Times*, January 24, 1933.

187. "Letter to the Prussian Academy of Sciences." *Science*, new series, 77 (1933): 444.

In this letter of April 5, 1933, Einstein stated his reasons for resigning from the Prussian Academy of Sciences (he had resigned in a letter of March 28): that he did not wish to live "in a country where the individual does not enjoy equality before the law, and freedom of speech and teaching." He also expressed the opinion that the academy had slandered him by accusing him of "atrocity-mongering" against Germany in America and France.

188. "Victim of Misunderstanding." *Times* (London), September 16, 1933.

In this letter Einstein explained his position on communism.

189. "Civilization and Science." Speech presented at Royal Albert Hall, London, October 4, 1933. Published as "Europe's Danger, Europe's Hope" in 1934 by Friends of Europe Publications, no. 4.

The meeting was organized by the Refugee Assistance Fund. Einstein spoke on the interrelationship between personal freedom and collective security and concluded that "only through peril and upheaval can nations be brought to further development." The *New York Herald Tribune* reprinted the speech as "Personal Liberty" on February 4, 1934.

190. "Dirac Equations for Semivectors" (Dirac-Gleichungen für Semi-Vektoren) (with Walther Mayer). *Akademie van wetenschappen* (Amsterdam). *Proceedings* 36 (1933): 497–502.

191. "Splitting of the Most Natural Field Equations for Semivectors into Spinor Equations of the Dirac Type" (Spaltung der natürlichsten Feldgleichungen für Semi-Vektoren in Spinorgleichungen vom Diracschen Typus) (with Walther Mayer). *Akademie van wetenschappen* (Amsterdam). *Proceedings* 36 (1933): 615–619.

192. "On the Method of Theoretical Physics." Herbert Spencer Lecture at Oxford University, June 10, 1933. Oxford: Clarendon, 1933. Also published in *The World as I See It* (1934) and *Ideas and Opinions* (1954).

In this lecture series in honor of philosopher Herbert Spencer, Einstein spoke about the development of the theoretical system, "something ineffable about the real, something occasionally described as mysterious and awe-inspiring," and the function of pure reason in science. He maintained that pure thought can grasp reality, using mathematical concepts to justify his confidence.

193. "Notes on the Origin of the General Theory of Relativity." George A. Gibson Foundation Lecture at Glasgow University, June 20, 1933. Glasgow University Publications No. 20. Glasgow: Jackson, 1933. Reprinted in *Ideas and Opinions* (1954).

The sponsors of the lecture asked Einstein to speak about the history of his own scientific work. He agreed to do so because, as he said, it is easier to throw light on one's own work than on someone else's and one should not neglect to do so out of modesty. He traced the work of others who had influenced him and which eventually led to his discoveries, as well as outlining the obstacles he had to overcome in his thinking.

1934

In America, the Einsteins immediately involved themselves in the intellectual and political activities that had always been of concern and interest to them. One of their first invitations came from President Franklin D. Roosevelt, with whom they dined in the White House in late January, staying overnight there. Even here, according to Helen Dukas, Einstein did not wear socks.

One of Einstein's first acts in America was to give a benefit violin recital in New York as a fund-raiser for scientists who were fleeing Germany.

Einstein's financial security was now assured. He had had some money put away in foreign accounts, and he received a salary twice that of the average university professor—a total luxury during the Great Depression. Also, his lecture fees were high. Even though his own needs were minimal, Elsa enjoyed the amenities of a bourgeois lifestyle.

Einstein was still interested in pacifism in general, and he decided that the best approach to peace was to speak out in favor of a world government, which he felt was the best defense against fascism.

In May, the family received the bad news that Ilse was deathly ill in Paris. Elsa rushed to Europe by ship to be at her bedside, only to watch Ilse die in July at the age of thirty-seven. Meanwhile, Einstein spent a long summer on the shores of Rhode Island, along with a physician friend from Berlin, Gustav Bucky, and his family, sailing and enjoying the sea while Elsa was away. Later that year, Margot and Dmitri came to Princeton, while the widower Rudolf Kayser, Ilse's husband, remained in Europe and settled in the Netherlands.

In June, Hitler, as Stalin had done in the USSR, began to remove and assassinate much of the political and military opposition in Germany and purge the Nazi Party of those he didn't trust. Because of an alleged plot by some Nazis against Hitler and his government, he had a number of Nazis executed. After German president Paul von Hindenburg died, Hitler installed himself as Führer, demanding the loyalty of all civilians and the military. The salute and cry of "Heil Hitler," already compulsory among Nazi Party members, became a norm in German society. To the southeast, Austrian chancellor Engelbert Dollfuss was assassinated by the Nazis after he had ruthlessly suppressed a national-socialist uprising in the country, but a Nazi attempt to take over Austria failed.

MARIE CURIE DIED, and her daughter Irène now took the spotlight. Together with Frédéric Joliot, she announced the discovery of artificial radiation obtained by bombarding certain nuclei with alpha particles. Later that year, Enrico Fermi and colleagues used neutrons to bombard uranium. They established that slow neutrons (which had passed through paraffin) were more efficient

Bohr (right) and Heisenberg at a Niels Bohr Institute conference.

Einstein plays violin with Paul Ehrenfest on the piano.

than fast ones in producing certain nuclear reactions and showed how nuclear reactions could be controlled.

Leo Szilard filed the first patent application for the idea of a neutron chain reaction. He assigned it to the British admiralty the next year to keep the patent a secret.

John Wheeler and Gregory Breit calculated the probability that two colliding photons could create an electron-positron pair. Their work was confirmed sixty-four years later at the Stanford Linear Accelerator.

Edwin Hubble and Milton Humason established photographically that there were at least as many galaxies in the universe as there are stars in the Milky Way. And James Chadwick and Maurice Goldhaber made the first determination of the neutron mass that was accurate enough to decide that the neutron was indeed a fundamental building block.

The Nobel Prize money in the physics section was allocated to the Main Fund and to the Special Fund of this division. The chemistry prize went to Harold Urey of the United States for his discovery of heavy hydrogen (deuterium).

194. ***The World as I See It* (Mein Weltbild). New York: Covici-Friede, 1934.**

This book is a collection of excerpts and essays on a variety of topics; it is the first of several future editions. Later editions were abridged and do not contain some of the material in this original version. Among other essays, it includes Einstein's lectures at King's College in London and at Columbia University in New York, both in 1921, and various writings on science, Judaism, and politics published until that time. Many of these essays and lectures are reprinted in *Ideas and Opinions* (1954).

195. **"Education and World Peace." *Progressive Education* 11 (1934): 440. Reprinted in the *New York Times*, November 24, 1934.**

In this message read at a New York conference of the Progressive Education Association on November 23, 1934, Einstein said that the United States is in the fortunate position of teaching pacifism in the schools because, due to "no serious danger" of a foreign invasion, it is not necessary to inculcate a military spirit in students. He called for international rather than national military means of defense and a strengthening of international solidarity.

196. "Presentation of Semivectors as Vectors of a Particular Differentiation Character" (Darstellung der Semi-Vektoren als gewöhnliche Vektoren von besonderem Differentionscharakter) (with Walther Mayer). *Annals of Mathematics*, ser. 2, 35 (1934): 104–110.

197. "Introduction" (in English and German). In Leopold Infeld, *The World in Modern Science*. London: Gollancz, 1934.

198. "Obituary for Paul Ehrenfest" (Nachruf Paul Ehrenfest). In *Almanak van het Leidsche Studencorps*. Leiden: Doesburg, 1934.

Ehrenfest, born in Vienna, had been professor of theoretical physics at the University of Leiden and had had a very close friendship with Einstein since 1912. For many years, he had tried to attract Einstein to a permanent position in the Netherlands. Suffering from a "morbid lack of self-confidence," according to Einstein, he mercy-killed his sixteen-year-old son, Vassik, who suffered from Down's syndrome, and then shot himself.

1935

A house became available for the Einsteins at 112 Mercer Street, less than a mile's walk to both Fine Hall on the university campus and the new grounds of the Institute for Advanced Study, which would be Einstein's professional home until his death.

Though the Einsteins had come to the United States with only visitors' visas, their goal was to become U.S. citizens. At that time, an application for citizenship could be filed only at a foreign consulate in a foreign country. They chose to do so in Bermuda and took a cruise there in May. After filing their application, they were the guests of the American consul at a lavish party in Einstein's honor. Five years later, the required waiting time, Einstein would become a U.S. citizen.

A "snake oil" salesman putting on a show in Tennessee. Such demonstrations were still popular through the late 1930s.

Einstein also traveled to Massachusetts to receive an honorary degree from Harvard University. The Einsteins spent the summer in Old Lyme, Connecticut, a historic seaside village on the east bank of the mouth of the Connecticut River.

In Europe, Germany incorporated the Saarland, since 1919 an autonomous region in what is now southwestern Germany, into its Reich. The Nazis repudiated the

Einstein circa 1935

Versailles Treaty and stopped making World War I reparations while also reintroducing compulsory military service. As Germany began to prosper again, the Nazis passed the Nuremberg Laws, by which Jews lost all protections and security as German citizens, relegating them to second-class citizenship and prohibiting them from marrying or having sexual relations with "Aryan" Germans.

A PUNCH-CARD MACHINE introduced by IBM had an arithmetic unit based on relays and could perform multiplication.

Hideki Yukawa of Japan proposed that the force holding the atomic nucleus together resulted from the exchange of a new particle that was several hundred times heavier than an electron. Two years later it was established that Yukawa's particle was the π-meson, or pion.

Sir James Chadwick of England was awarded the Nobel Prize in physics for the discovery of the neutron. The chemistry prize was awarded jointly to Frédéric Joliot and Irène Joliot-Curie of France in recognition of their synthesis of new radioactive elements, or artificial radioactivity.

199. "Appeal for Jewish Unity." Address before the women's division of the American Jewish Congress. *New Palestine* 25, no. 9 (March 1, 1935): 1.

200. "Peace Must Be Waged." Interview by R. M. Bartlett. *Survey Graphic* 24 (1935): 384. Einstein offered his view on war and peace in this interview, dealing mainly with Germany's nationalism since the First World War. He said that every country needs to surrender a portion of its sovereignty through international cooperation, that people need to think in international terms, and that, to avoid destruction, humans must sacrifice aggression. He also stated his belief that humans can abolish war through education.

201. "Can Quantum-Mechanical Description Be Considered Complete?" (with B. Podolsky and N. Rosen). *Physical Review*, ser. 2, 47 (1935): 777–780.

Here the authors brought to public attention Einstein's critical attitude toward quantum theory ("spooky action-at-a-distance"). They elegantly exposed the consequences of the quantum-mechanical formalism regarding the representation of a state of a system that consists of two particles that have been in interaction for a limited time interval. According to their criteria, quantum mechanics does not provide a complete description of physical reality. If it *is* to be complete, then so-called hidden variables must exist. This paper later created contention among physicists, after the work of John Bell in 1964 propelled it into significance. Bell showed that quantum mechanics must be nonlocal. Subsequent experiments have confirmed Bell's theorem.

Nathan Rosen was a young student and collaborator of Einstein's who later became a founding member of the physics department at the Technion in Haifa, Israel. According to Abraham Pais, the idea for the paper came from Rosen. Before coming to the Institute for Advanced Study, Boris Podolsky had worked at Caltech in Pasadena. He wrote the actual paper after having long discussions with Einstein.

202. "The Particle Problem in the General Theory of Relativity" (with N. Rosen). *Physical Review*, ser. 2, 48 (1935): 73–77.

203. "Elementary Derivation of the Equivalence of Mass and Energy." J. W. Gibbs Lecture to the American Association for the Advancement of Science, December 28, 1934. *Bulletin of the American Mathematical Society* 41 (1935): 223–230.

Hitler preaches the superiority of the Aryan race, but African American Jesse Owens wins the gold at the 1936 Berlin Olympics.

1936

Unfortunately, Elsa did not have a chance to enjoy her new home and furnishings for long. She became ill with kidney disease and had a painful year, some of it spent at Saranac Lake in upstate New York. She died right before Christmas. Though Einstein was depressed during the time she was ill, he recovered rather quickly after her death. It may have helped that he was not alone in his household: he still had the company of Margot, who had gotten a divorce from Dmitri, and Helen Dukas—and his work. Three years later, his sister, Maja, would join them as well.

Hans Albert received a doctorate from the ETH (formerly the Poly) in Zurich. The following year he came to the United States to visit his father, and in 1938 he resigned his position in Switzerland and moved to the United States with his family. After spending some time in Greenville, South Carolina, working at the U.S. Department of Agriculture, and at Caltech, he became a professor of hydraulic engineering at the University of

California in Berkeley in 1947. After his wife, Frieda, died in 1958, he married Elizabeth Roboz.

Albert Einstein came to understand and read English quite well, but the middle-aged man never mastered spoken English and in letters to friends often complained about his difficulties. He was able to write and speak simple sentences, and he could read prepared speeches, but in general he reverted to German in long discussions or conversations. Many of his letters at this time were first written in German and then translated into English by Helen Dukas or colleagues.

German troops occupied the Rhineland, a western area near the French border that had been under weak French control since the Treaty of Versailles. They remilitarized the area without opposition. Hitler's foreign policy and plan for domestic construction were approved by an astounding 98.8 percent of the voters in a new German referendum on March 29. His influence all over Europe grew as pro-Nazi groups wanted their countries to become part of the Third Reich, which Hitler was determined to expand. Many countries and autonomous regions were easy targets as they were going through the Great Depression and Hitler promised to restore Germany's power and prestige, lost after its defeat in World War I. Not all were eager to join, however. An Austro-German convention reiterated that Austria was an independent country, a state of affairs that would change in two years with the *Anschluss* (union of the two countries).

IN MEDICINE AND THE SCIENCES, great strides were being made. Dr. Alexis Carrel and his lab assistant, the aviator Charles Lindbergh, developed the first artificial heart to be used during heart surgery, by inventing the perfusion pump.

Edwin Hubble, in his book *The Realm of the Nebulae,* suggested that the universe extends out about 500 million light-years. This distance has been revised upward several times since then, making our own galaxy just a tiny speck in a vast cosmos.

English mathematician Alan Turing, while a graduate student at Princeton University, published a paper in which he described the "Turing machine," the abstract precursor of the computer. Turing committed suicide in 1954 in England by cyanide poisoning.

Hans von Ohain, who designed the first operational turbojet engine, gained a patent for his machine. Frank Whittle had already filed the first patent for a jet engine in 1930, but Von Ohain's jet was the first to fly, in 1939.

The first television broadcast took place in England.

The Nobel Prize in physics was divided equally between Victor F. Hess of Austria for his discovery of cosmic rays and Carl D. Anderson of the United States for his discovery of the positron. The chemistry prize was awarded to Peter Debye of the Netherlands (though he worked in Berlin at the time of the award) for his contributions to our knowledge of molecular structure through investigations on dipole moments and on the diffraction of x-rays and electrons in gases.

204. "Some Thoughts Concerning Education." Translated by Linda Arronet. *School and Society* 44 (1936): 589–592. Reprinted in *Ideas and Opinions* (1954) as "On Education."

In this speech, given at a convocation of the State University of New York at Albany to celebrate the tercentenary of higher education in America, Einstein said that the aim of education must be to train independently thinking individuals whose highest life goal should be to serve their community. Schools should not use fear, force, and artificial authority—all of which he had detested in the German educational system of his youth.

205. "Freedom of Learning." *Science* 83, new ser. (1936): 372–373.

206. "Physics and Reality." *Journal of the Franklin Institute* 221, no. 3 (March 1936): 313–347. Reprinted in *Ideas and Opinions* (1954).

Einstein, who was awarded the Franklin Medal in 1935, argued that the quantum-mechanical description can be considered only as a way of accounting for the average behavior of a large number of atomic systems. In the following words, he expressed his belief that it should give exhaustive descriptions of individual phenomena: "To believe this is logically possible without contradiction, but it is so very contrary to my scientific instinct that I cannot forgo the search for a more complete conception."

207. "The Two-Body Problem in General Relativity" (with N. Rosen). *Physical Review*, ser. 2, 49 (1936): 404–405.

208. "Lens-like Action of a Star by Deviation of Light in the Gravitational Field." *Science* 84 (1936): 506–507.

1937

Though Einstein continued to give speeches and interviews and carried on a hefty correspondence, he published only one paper this year, and this one was in collaboration with someone else. In fact, his scientific work now consisted almost entirely of collaborations with young assistants as his working energy declined.

Einstein continued to help refugees come to the United States by trying to find them sponsors or jobs and lending or giving them money when necessary.

USING A CLOUD CHAMBER, J. C. Street and E. C. Stevenson discovered the muon. Before this, scientists had thought that electrons, protons, and the newly discovered neutron were the only fundamental particles. The muon is an unstable lepton (a particle that has spin quantum number ½ and experiences no strong forces) that is common in the cosmic radiation near Earth's surface.

Grote Reber constructed the first radio telescope in his backyard in Wheaton, Illinois. He built the telescope at his own expense while working for a radio company in Chicago.

Sir Ernest Rutherford, discoverer of the nucleus, dies.

The Nobel Prize in physics was awarded jointly to Clinton Joseph Davisson of the United States and Sir George Paget Thomson of England for their experimental discovery of the diffraction of electrons by crystals. The chemistry prize was divided equally between Sir

After four years and $35 million, San Francisco's Golden Gate Bridge opens for traffic.

Walter N. Haworth of England, for his investigations on carbohydrates and vitamin C, and Paul Karrer of Switzerland, for his investigations on carotenoids, flavins, and vitamins A and B2.

209. "On Gravitational Waves" (with N. Rosen). *Journal of the Franklin Institute* 223 (1937): 43–54.

1938

This year, great waves of immigration into the United States followed the annexation *(Anschluss)* of Austria by Germany in March. Einstein could no longer handle the flow of appeals that came to him, and his finances became limited while America was still going through the Depression. He was now part of an ever-growing number of Jewish refugees in America, many of them academics, though he did not socialize with them exclusively. His closest friends were educated, nonacademic professionals such as physicians, artists, and writers.

Einstein's second grandson, Hans Albert and Frieda's son Klaus, died suddenly late in the year at the age of six, possibly of diphtheria. Einstein wrote to the couple in January 1939, after he had learned of the death: "The deepest sorrow loving parents can experience has come upon you... Although I saw [Klaus] only for a short time, he was as close to me as if he had grown up near me."

Sensing a tumultuous time ahead, President Roosevelt sent an appeal to Germany and Italy to settle European problems amicably, but the United States withdrew its ambassador to Germany, and Germany, not to be outdone, recalled its ambassador to the United States. The infamous *Kristallnacht* took place in November, and Hermann Goering, the outrageous, brutal, and cunning commander in chief of the German Air Force who had gained power in economic and political circles, fined Germany's Jews a million marks and ordered the "Aryanization" of all Jewish businesses. Goering used his position to indulge in ostentatious luxury. He occupied a palace in Berlin and built a lavish hunting lodge where he organized feasts and hunting expeditions and in general showed off his extravagant tastes. Charles Lindbergh had accepted the German Medal of Honor from Goering, causing outrage in the United States. The following year he made himself even more unpopular by criticizing FDR's policies, whereupon the president denounced Lindbergh. The aviator never returned the medal.

Enrico Fermi receives the Nobel Prize for his work on artificial radioactivity.

THIS YEAR IN PHYSICS, Otto Hahn, Lise Meitner, Fritz Strassmann, and Meitner's nephew Otto Frisch discovered nuclear fission, which had revolutionary effects not only in physics but later also on world politics, when nuclear weapons became an issue. Hans Bethe hypothesized that nuclear fusion is the source of energy in stars, and Arthur Compton demonstrated that cosmic radiation consists of charged particles.

The Nobel Prize in physics was awarded to Italy's Enrico Fermi for his demonstration of the existence of new radioactive elements produced by neutron irradiation and for his related discovery of nuclear reactions brought about by slow neutrons. Germany's Richard Kuhn received the chemistry prize for his work on carotenoids and vitamins. Kuhn was pressured not to accept the award because Hitler had raged against the Nobel committee after it had awarded the peace prize in 1935 to Carl von Ossietzky, a Nazi opponent. The German government informed Kuhn that accepting the prize would not be in the best interests of the country, forcing him to forgo the prize money. Later, he did at least accept the medal and certificate.

210. "Our Debt to Zionism." Address to National Labor Committee for Palestine. *New Palestine* 28, no. 2 (April 29, 1938): 2–4. Reprinted in *Ideas and Opinions* (1954).

This essay is part of an address given to the National Labor Committee for Palestine in New York City on April 17, 1938. Focusing on the current troubled times for Jews, Einstein said that Zionism had renewed a sense of community among the Jews, enabling many of them to escape anti-Semitism and engaging them in productive work in Palestine.

211. "Why Do They Hate the Jews?" Translated by Ruth Norden. *Collier's Weekly* 102 (November 26, 1938): 9–10, 38. Reprinted in *Ideas and Opinions* (1954).

In answer to this question, Einstein said that Jews were the object of discrimination because they are thinly scattered throughout the Diaspora and therefore unable to defend themselves against attacks, usually initiated by envious antagonists. Historically, they have been charged both with trying to assimilate and with being too clannish. He maintained that societies need heterogeneous groups, both political and social, for their invigorating effect on all aspects of life.

212. "Gravitational Equations and the Problems of Motion" (with Leopold Infeld and Banesh Hoffmann), part 1 (see paper 223 for part 2). *Annals of Mathematics* 39, ser. 2 (1938): 65–100.

Discussing their theory of the interaction of point masses with gravity, the three physicists showed that the laws of motion of such particles follow the gravitational field equations.

213. "Generalization of Kaluza's Theory of Electricity" (with Peter Bergmann). *Annals of Mathematics* 39, ser. 2 (1938): 683–701.

Peter Bergmann was Einstein's assistant at the Institute for Advanced Study until 1941, and they coauthored two papers: this one and one in 1941, which included another assistant, Valentine Bargmann, as a third author. In 1942, Einstein wrote the foreword for Bergmann's textbook, *Introduction to the Theory of Relativity.*

214. *The Evolution of Physics: The Growth of Ideas from Early Concepts to Relativity and Quanta* (with Leopold Infeld). New York: Simon and Schuster, 1938.

In this popular book, Einstein and Infeld presented the ideas behind relativity theory and traced physical thought since the time of Galileo. They did so without the use of mathematics, making this a useful guide for laypeople and physicists alike. Infeld, who had come to the United States from Poland, worked as an assistant to Einstein at the Institute for Advanced Study from 1936 to 1938. From 1938 to 1950 he worked at the University of Toronto, but then the Canadian government accused him of dispensing secret atomic bomb information to the Soviet Union and he returned to Poland, dying there in 1968.

1939

Einstein's sister, Maja, came to Princeton from Italy. Her husband, Paul Winteler, had been refused entry into the United States because of health problems, which at that time made one ineligible for immigration. He moved to Geneva to live with friends, but Maja decided to move to Princeton to be in a safer haven with her brother, hoping to rejoin Paul later. The Einstein household on Mercer Street now consisted of Einstein, Maja, Margot, Helen Dukas, and a dog, Chico, and cat, Tiger. All of them would continue to live in the house until their deaths, with Margot being the last to die in 1986.

By now Einstein was becoming an outsider in the world of physics, no longer involved with mainstream research. Still, he kept abreast of the newest developments, debated quantum mechanics, and worked on his

Home sweet home—Einstein's house on Mercer Street in Princeton

unified field theory. He gained scientific satisfaction vicariously through his younger associates and continued the humanitarian and political involvements he championed.

Because of his reputation as a genius and humanitarian, Einstein's name continued to carry a lot of weight, and his involvement in a project often led to its success. On a muggy July day, his old friend Leo Szilard (from the refrigerator days) visited Einstein's summer retreat at Peconic Bay on Long Island. Another physicist, Eugene Wigner, was at his side. They reported that, in new experiments with uranium, German chemists Otto Hahn and Fritz Strassmann had demonstrated the possibility of splitting the atom of uranium-235 (which, as Austrian physicist Lise Meitner explained to them from her exile in Sweden, was "nuclear fission"). With this development, scientists would soon be able to achieve a chain reaction that could produce a huge amount of energy—possibly a fission bomb—and the visitors feared that Germany might begin work on such a bomb. They reported that, recognizing the implications, Frédéric Joliot of France had immediately ordered six tons of uranium oxide from the Belgian Congo and heavy water (used as a moderator and coolant in nuclear reactors) from Norway and sent the material to England ahead of the invading German army. The visitors and their host anxiously discussed the scenario that Germany might also recognize the possibilities inherent in chain reactions and embark on an atomic bomb project of its own.

By August, Szilard had drafted a letter to President Roosevelt, which Einstein was to sign, warning the president about the military implications of atomic energy. The letter stressed that a catastrophe might lie ahead if Germany succeeded in building a bomb and the United States neglected to do so. Einstein signed the letter, and it was delivered to Roosevelt—though not until more than a month after World War II had already begun on September 3. Roosevelt thanked Einstein for the important news and proceeded to appoint a committee to do a year's work of research on uranium and to study the possibilities of nuclear energy.

The World's Fair took place in New York, and Einstein and the members of his household took time to enjoy the exhibits. Einstein submitted a statement, "To Posterity," for the fair's time capsule. In it, he recalled some of the world's technological achievements and informed future generations that people were living in both economic fear and fear of war. He said that the intelligence and character of the masses are lower than those of the few people who had achieved something that was of value to the world.

Einstein was present when the cornerstone of Fuld Hall, the Institute for Advanced Study's first building, was laid on May 22.

On September 1, Germany invaded Poland and annexed the northern German-speaking port city of Danzig (Gdansk). Within two days, Britain and France declared war on Germany, while the United States remained neutral. Germany continued to overrun western Poland and took Warsaw, the capital city, immediately installing a Nazi governor-general in Poland. At the same time, the USSR was invading Poland from the east. In England, women and children were evacuated from London—where the English translation of Hitler's

Einstein and family members at the World's Fair in New York

Mein Kampf was now being published—in anticipation of war. In the United States, the economy began to recover as the European nations ordered arms and materiel in preparation for war. Meanwhile, at the dawn of World War II in Europe, Sigmund Freud died in London at the age of eighty-three.

IN 1939, NIELS BOHR proposed his "droplet model" of the nucleus, which paved the way for nuclear fission. According to this model, short-range forces pull nuclear particles together, similar to the way in which molecules of a liquid attract one another. Basically, the theory stated that a fission reaction might be initiated if a neutron hit the heavy nucleus of an atom. Bohr discovered that the rare uranium-235 was fissionable and the more common isotope uranium-238 was not, which became important to the work of the Manhattan Project.

The ever-inventive Szilard proposed stacking alternate layers of graphite and uranium in a lattice; this geometry would define neutron scattering and subsequent fission events. I. I. Rabi and his collaborators developed a method for measuring nuclear magnetic moments, which forms part of the basis for lasers, atomic clocks, and measurement of the Lamb shift. Americans viewed their first television broadcast, launching a sea change in home entertainment.

Ernest O. Lawrence of the United States was awarded the Nobel Prize in physics for the invention and development of the cyclotron and for results obtained with it, especially with regard to artificial radioactive elements. The chemistry award was given to Adolf Butenandt of Germany for his work on sex hormones and to Leopold Ruzicka of Switzerland for his studies on polymethylines and higher terpenes. As with Richard Kuhn the preceding year, the German government pressured Butenandt into refusing to accept his prize on political grounds, but, like Kuhn, he received the medal and certificate later.

Albert Einstein circa 1939

215. "Our Goal." Address at a conference at Princeton Theological Seminary, May 19, 1939. Mimeograph. Reprinted as part 1 of "Science and Religion" in *Ideas and Opinions*, 41–44. See paper 225 for part 2. Also see papers 156 and 256 on the same topic.

According to Einstein, scientific and rational means cannot fully serve to influence a person's convictions and beliefs—they have their limits. Therefore, the most important function of religion in our social lives is to make clear a society's values and goals, its powerful traditions that have given a foundation to aspirations and values. Traditions do not need to be justified because they have already worked well in a healthy society.

216. [Sixtieth Birthday Statement.] *Science* 89, new ser. (1939): 242.
Einstein very briefly discussed the American scientific spirit, expressing appreciation for the ideal working and living conditions he was enjoying in the United States.

217. "Europe Will Become a Barren Waste." *New Palestine* 29 (March 24, 1939): 1–2.
Based on a radio address in support of the United Jewish Appeal entitled "Humanity on Trial," this statement was also distributed as a leaflet.

218. "Stationary Systems with Spherical Symmetry Consisting of Many Gravitating Masses." *Annals of Mathematics*, ser. 2, 40 (1939): 922–936.

1940

British prime minister Winston Churchill makes a wartime radio address.

Noting that no decisive action had been taken after his warning about atomic energy, Einstein sent a second letter to President Roosevelt in March, stressing the urgency of the situation. But the pace did not pick up until the fall of 1941.

Einstein became an American citizen on October 1. Along with Helen Dukas and stepdaughter Margot, he went to the courthouse in Trenton to take his oath of allegiance. Einstein retained his Swiss citizenship, however, and held dual citizenship until the end of his life.

In the midst of the war in Europe, under skies heavy with military planes, Winston Churchill became prime minister of Britain after Neville Chamberlain resigned; Churchill passionately delivered his famous "blood, sweat, and tears" speech to rally his country and prepare them for what lay ahead. Germany invaded Norway, Denmark, Holland, Belgium, and Luxembourg; the Dutch and Belgians surrendered and German occupation began in the two countries. Italy declared war on

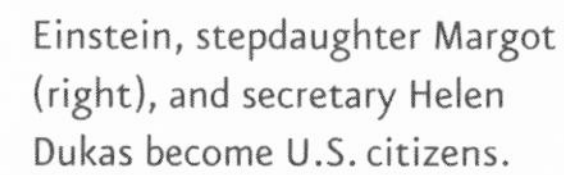

Einstein, stepdaughter Margot (right), and secretary Helen Dukas become U.S. citizens.

France and Britain, and the Germans entered Paris in June. Germany conducted a *Blitzkrieg* (literally, lightning war—here, all-night, intensive aerial bombing) of London. Japan, Germany, and Italy—the Axis Powers—signed a military and economic pact. Throughout all of this turmoil, many of Europe's greatest scientists managed to flee to the United States.

In America, voters elected Franklin Roosevelt as president for an unprecedented third term. In anticipation of war, he asked Congress for a huge defense budget and urged the production of fifty thousand airplanes. Congress created the Selective Service System for compulsory military service, and all men between the ages of twenty-one and thirty-six were required to register immediately. The Smith Act required all aliens to register.

BY MARCH 1940, physicists Otto Frisch and Rudolf Peierls in England had produced two brief but significant papers showing that, if the rare uranium-235 isotope could be separated, the amount needed for a bomb would be relatively small and therefore the building of a bomb would be feasible.

Georgii Flerov and Konstantin Petrzhak of the Soviet Union discovered the spontaneous fission of uranium. Harold Urey became director of the U.S. program to separate uranium isotopes; while doing so, he developed statistical methods of isotope separation that permitted large-scale production of uranium-235. Further, Norbert Wiener proposed the construction of vacuum-tube electronic computers that could, by using binary mathematics on magnetic tape, make totally preprogrammed digital calculations.

Vannevar Bush, director of the Carnegie Institution of Washington, organized the National Defense Research Committee to mobilize America's scientific resources in support of the war effort.

This year through 1942, no Nobel Prizes were awarded in physics and chemistry. Instead, the prize money for these years was allocated to the foundation's Main Fund and to the Special Fund of each section.

219. "Freedom and Science." Translated by James Gutmann. In *Freedom: Its Meaning*, edited by Ruth N. Anshen, 381–383. New York: Harcourt, Brace.

After stating that most people can agree on two goals—to satisfy their basic physical needs through the least amount of work possible and to be spiritually content, for which they must have the chance to develop their intellectual and artistic gifts—Einstein said that the freedom to express oneself is paramount. People should not have to work so much that they have neither the time nor the strength to express themselves creatively or with independent ideas.

220. "New Bond among Nations" (Neuer Bund der Nationen). *Aufbau* 6, no. 26 (December 27, 1940): 1–2.

This article is based on an interview at the time Einstein received his American citizenship.

221. "My Position on the Jewish Question" (Meine Stellung zur jüdischen Frage). *Aufbau* 6, no. 52 (December 27, 1940): 9.

222. [Statement on the significance of American citizenship.] In *I Am an American: By Famous Naturalized Americans*, edited by Robert Spears Benjamin, 43–47. New York: Alliance Book.

223. "Gravitational Equations and the Problems of Motion" (with Leopold Infeld), part 2 (see no. 212 for part 1). *Annals of Mathematics* 41, ser. 2 (1940): 455–464.

224. "Considerations Concerning the Fundamentals of Theoretical Physics." *Science* 91, new series (1940): 487–492. Reprinted in *Ideas and Opinions* (1954) as "The Fundaments of Theoretical Physics." Address to the Eighth American Scientific Congress, Washington, D.C., May 24, 1940.

Einstein traced the development of theoretical scientific thinking and the attempt to find a unifying theoretical basis for each branch of science.

1941

The destruction at Pearl Harbor, as photographed by a Japanese pilot

Einstein became a grandfather again, with the adoption of Evelyn, born in March, by Hans Albert and Frieda.

On the eve of the attack on Pearl Harbor, the U.S. government at last decided to embark on a massive technological and scientific enterprise in support of the development of atomic energy. Plans were made for the highly classified Manhattan Project, to be carried out by eminently skilled physicists and technical experts in laboratories in Los Alamos in the New Mexico desert. When the Manhattan Project was established, its first major challenge was to find acceptable and plentiful sources of fuel for the bombs. Both uranium-235 and plutonium-239 were likely candidates because they could sustain a chain reaction, which gives the atomic bomb its power.

Einstein was ready to help in the effort but was unable to do so: he could not get the required security clearance. His newly adopted homeland considered him a security risk because of his leftist sympathies. Indeed, agents had been following his every move for years and were continuing to do so without his knowledge. The massive FBI file on Einstein can now be seen on the FBI's Website on the Internet and is discussed thoroughly in a recent book by Fred Jerome. In the end, the FBI reports concluded that Einstein did not appear to have Communist leanings, nor was he involved in subversive activities. Still, his fearlessness in associating socially with Soviet sympathizers made him a suspect citizen. Had FBI agents known that Einstein's warnings helped to establish the Manhattan Project to defend the nation, they might have been more likely to consider him a loyal American.

In later years, after witnessing the devastation and agony caused by atomic weapons, Einstein said if he had known that the Germans would not be successful in building a bomb, he would never have signed the letter to Roosevelt. Though German scientists had been aware of the possibilities, they were not able to build a bomb because the necessary resources were not available to them under war conditions. Still, a nuclear reactor running at criticality was almost achieved by the Germans by the end of the war.

The U.S. government did seek Einstein's advice over the next few years on nonclassified matters and on explosives. He felt useful in being part of the effort to defeat fascism while being spared from any direct involve-

The carving of Mount Rushmore was halted in 1941 when the United States entered World War II.

Swing sweeps the nation. Count Basie and his orchestra are among the most popular of the big bands.

ment in the development of the most destructive weapon ever built.

World War II was already in full swing all over Europe, and Germany opened its counteroffensive in North Africa and invaded Russia, seeking to expand the Third Reich. The German battleship *Bismarck* sunk the British battleship HMS *Hood* after one of the fiercest naval battles in history, and then Allied torpedoes succeeded in sinking the German vessel; the wreck of the *Bismarck* was discovered in 1989 at a depth of 15,700 feet, and the *Hood* was found in 2001.

After the U.S. ambassador to Japan warned Roosevelt of a possible Japanese attack on the United States and Pearl Harbor was hit on December 7, the United States and Britain declared war on Japan. On December 10, Germany declared war on the United States. Italy followed suit, and the United States declared war on both countries. In the U.S. Congress, only one congressional representative, pacifist Jeanette Rankin of Montana, cast a vote against declaring war.

GLENN SEABORG and his colleagues at the University of California in Berkeley discovered the element plutonium. One of its isotopes (P-239, which they produced from P-238) was fissionable and suitable for an atomic bomb. Its fission rate was even greater than that of uranium-235.

Lev Landau of the Soviet Union constructed a complete theory of quantum liquids at very low temperatures. He would win the Nobel Prize in physics in 1962.

No Nobel Prizes in physics and chemistry were awarded this year.

225. "Science and Religion, Part 2." In *Science, Philosophy and Religion*, a symposium volume published by the Conference on Science, Philosophy and Religion in Their Democratic Way of Life, New York, 1941. Reprinted as part 2 of "Science and Religion" in *Ideas and Opinions* (1954), 44–49. See paper 215 for part 1. Also see papers 156 and 256 on the same topic.

Einstein said that he can define science, but he cannot define religion. Therefore, if a person claims to be religious, he assumes it is someone who has values of a superpersonal, not materialistic, nature. Conflicts between science and religion have sprung from fatal errors, religion can learn from science to achieve its goals, and the source of the need to seek truth through science springs from the religious sphere, that is, from a faith that the rules valid for the world of existence are rational. He made his famous remark, "Science without religion is lame, religion without science is blind."

226. "Five-Dimensional Representation of Gravitation and Electricity" (with V. Bargmann and P. G. Bergmann). In *Theodore von Karman Anniversary Volume*, 212–225. Pasadena, Calif.: California Institute of Technology, 1941.

227. "Credo as a Jew." In *Universal Jewish Encyclopedia* 4 (1941): 32–33.

228. "The Common Language of Science." *Advancement of Science* (London): 2, no. 5 (1941). Reprinted in *Ideas and Opinions* (1954).

In this broadcast recording for a science conference in London on September 28, 1941, Einstein spoke of language as an instrument of reasoning and of the intimate connection between language and thinking.

1942 1943

Einstein officially became a consultant on explosives for the U.S. Navy, for which he was paid twenty-five dollars a day.

Bertrand Russell arrived in Princeton, beginning a friendship with Einstein and meeting for discussions with him and others once a week.

The United States appointed General Douglas MacArthur commander in chief of the troops in the Far East and General Dwight D. Eisenhower in Europe and North Africa. The Japanese continued to invade and capture territories in the Pacific and defeated the U.S. fleet at the Battle of the Java Sea. The Japanese occupied Bataan on the Philippine Peninsula and forced seventy thousand American and Filipino soldiers to march more than sixty miles to a northern camp; about 10 percent died on the way, while the others were brutally mistreated and starved during what became known as the Bataan

Uneasy alliance: Josef Stalin, Franklin Delano Roosevelt, and Winston Churchill

"Women had a big share in filling these bundles for Berlin."

Death March. In Germany and Poland, the murder of millions of Jews, Slavs, and dissidents began in the Nazi death camps. Allied round-the-clock bombing of Germany began by the end of 1943.

In the United States, Japanese Americans of all ages were undergoing horrors of their own. Even those born in the United States or holding U.S. citizenship were now considered a threat by the government and were forced to leave their homes on the West Coast and in Latin America. Caravans of military vehicles transported them to ten special inland detention camps until the end of the war. A little-known fact is that many Italian Americans were forced to move as well, all this as both Japanese and Italian sons were sent to war on the American side. More than fifty years later, the Japanese would receive minor reparations for the indignities they suffered, while the Italians were still struggling to get their story known. In 2000, Congress asked the Civil Rights Division of the U.S. Department of Justice to investigate reports that thousands of Italian Americans were arrested, rounded up, or interned and that their homes were raided and property was confiscated during World War II.

WORK BEGAN at the Clinton Engineer Works (now Oak Ridge) in Tennessee to extract uranium-235 from uranium ore. Early in 1942, General Leslie Groves was appointed director of the whole Manhattan Project, and he in turn appointed J. Robert Oppenheimer as director of research at the Los Alamos National Laboratory, responsible for developing the atomic bomb.

In early 1942, physicist and Nobel laureate Arthur Holly Compton organized the Metallurgical Laboratory at the University of Chicago to study plutonium and fission piles. In December, Italian physicist and Nobel Prize winner Enrico Fermi, working at the University of Chicago's nuclear reactor, supervised the first controlled nuclear chain reaction, releasing controlled energy from the nucleus of the atom and laying the foundation for an atomic bomb. After Niels Bohr had updated him on the progress of fission, Fermi had immediately seen the possibility of the emission of neutrons at the start of a chain reaction—and when neutrons hit the heavy nucleus of an atom, a fission reaction could begin. Fermi continued to help solve the problems associated with the development of the bomb at the Manhattan Project; he became a U.S. citizen in 1944.

Meanwhile, John Atanasoff and Clifford Berry, at Iowa State University in 1942, finished building the first electronic digital computer, the Atanasoff-Berry Computer, or ABC. It incorporated several major innovations, including the use of binary arithmetic, regenerative

U.S. Marines are the first to desegregate. Blacks and whites share time on the diamond at Camp Lejeune.

Einstein works for the U.S. Navy.

memory, parallel processing, and the separation of memory and computing functions.

No Nobel Prizes in physics and chemistry were awarded in 1942. For 1943, the physics award went to Otto Stern of the Carnegie Institute of Technology in Pittsburgh for his contribution to the development of the molecular ray method and his discovery of the magnetic moment of the proton. The chemistry prize went to George de Hevesy of Hungary for his work on the use of isotopes as tracers in the study of chemical processes.

229. "The Common Language of Science." *Advancement of Science* 2, no. 5 (1942): 109.
A radio address to the British Association for the Advancement of Science, given in September 1941 but not published until the following year.

230. "Demonstration of the Non-existence of Gravitational Fields with a Non-vanishing Total Mass Free of Singularities." *Tucuman Universidad Nacional, Revista*, ser. A, 2 (1942): 11–15.

Based on an address before a joint meeting of the American Physical Society and the American Association of Physics Teachers, Princeton, N.J., December 29, 1941, entitled "Solutions of the Finite Mass of the Gravitational Equations."

231. "Nonexistence of Regular Stationary Solutions of Relativistic Field Equations" (with Wolfgang Pauli). *Annals of Mathematics* 44 (1943): 131–137.

1944

Einstein made a major contribution to the war effort this year. To raise money for war bonds, he had agreed to copy his 1905 paper on relativity by hand and allow it to be sold at auction. He contributed a second manuscript for the same cause. The manuscripts were auctioned in February: the relativity copy fetched $6.5 million, and the other manuscript $5 million. Einstein was astonished. The buyers donated both copies to the Library of Congress.

As the cost of living rose 3 percent in the United States, the war was still pummeling Europe, the Pacific, and North Africa, with gains and losses on all sides. In June, D-day (the Normandy invasion) took place—a huge Allied offensive on the French coast, with more than seven hundred ships and four thousand small craft meeting fierce German resistance and counterattacks.

In July, a growing number of disaffected German military officers who felt peace could be achieved only by Hitler's removal conspired in an attempt to assassinate the Führer. The effort failed when the briefcase containing a bomb was bumped aside from Hitler's path, and

Einstein and Bohr debate quantum mechanics.

Einstein with chorus girls

Hitler suffered only minor injury. German field marshall Erwin Rommel (the "Desert Fox") supported Hitler's removal but did not take part in the assassination plot; instead, he committed suicide later in the year, knowing that Hitler would get him eventually.

JOHN VON NEUMANN and Oskar Morgenstern published their formulation of game theory as it applies to human economic behavior, using zero-sum parlor games such as poker and chess. Stanislaw Ulam and Edward Teller, while working on the Manhattan Project, developed a two-stage radiation implosion design that employed both fusion and fission and allowed the detonation of thermonuclear weapons.

The Nobel Prize in physics was awarded to I. I. Rabi of the United States for his resonance method of recording the magnetic properties of atomic nuclei. The chemistry prize went to Otto Hahn of Germany for his discovery of fission of heavy nuclei.

232. **"Gandhi's Statesmanship." In *Mahatma Gandhi: Essays and Reflections on His Life and Work*, edited by S. Radhakrishnan. London: Allen and Unwin, 1944.**

This work was originally presented to Gandhi on his seventieth birthday in 1939 and then republished. Einstein was an ardent admirer of Gandhi and his nonviolent methods.

233. "The Arabs and Palestine" (with Erich Kahler). *Princeton Herald*, April 14 and 28, 1944.

Einstein vigorously stated his belief that a place must be made for the Arabs in Palestine.

234. "To the Heroes of the Battle of the Warsaw Ghetto." *Bulletin of the Society of Polish Jews* (New York) (1944).

In this two-paragraph statement, Einstein famously blamed "the Germans as an entire people" for the mass murders in Europe because they elected Hitler by a high majority while being aware of his intentions, as outlined in his book and speeches.

235. "Remarks on Bertrand Russell's Theory of Knowledge." In *The Philosophy of Bertrand Russell*, edited by Paul A. Schilpp. Library of Living Philosophers, vol. 5. LaSalle, Ill.: Open Court, 1944.

After stating his admiration for Russell, Einstein discussed the evolution of philosophic thought about the objective world and the world of concepts and ideas.

236. "The Ethical Imperative." *Opinion* 14 (March 1944): 10.

237. "Bivector Fields I" (with V. Bargmann). *Annals of Mathematics* 45, ser. 2 (1944): 1–14.

Valentine Bargmann, a mathematical physicist, was one of Einstein's assistants at the Institute for Advanced Study and a talented pianist. His wife, Sonja Bargmann, translated the essays in the popular book *Ideas and Opinions*, for which Bargmann wrote an introduction for the scientific part (pp. 217–220).

238. "Bivector Fields II." *Annals of Mathematics* 45, ser. 2 (1944): 15–23.

The United States drops atom bombs on Hiroshima and Nagasaki, ending World War II.

1945

Einstein continued to be concerned about the arms race and wrote President Roosevelt another letter in March at the request of Leo Szilard. He asked that the president speak with Szilard about his and other physicists' concerns about the lack of contact between those conducting classified research and those making policy. Einstein did not know the exact nature of the Manhattan Project, but presumably he was told that it could have grave consequences. Roosevelt, however, died on April 12 and never saw the letter. He also did not live to see the ending of German resistance that month, Hitler's suicide on April 30, and the official German surrender on May 7. The Allied offensive finally emerged victorious in Europe, and V-E Day (Victory over Europe) was declared on May 8. But Japan had not yet surrendered.

On July 16, an atomic bomb was tested at Alamogordo in the New Mexico desert. Three weeks later, on August 6, the U.S. B-52 bomber *Enola Gay* dropped a uranium-235 bomb, nicknamed "Little Boy," on Hiroshima. Three days later the bomber *Bock's Car* dropped a plutonium-239 bomb, "Fat Man," on Nagasaki. The shock waves, fire, and radiation in each attack claimed more than seventy thousand lives instantly, and thousands more died later. Japan surrendered and World War II ended on August 14, both in Europe and in the Pacific. Close to 50 million people had lost their lives by the end of the war, including those who had died in Nazi concentration camps.

The Smyth report at last disclosed that Einstein was the person who had warned Roosevelt about the possibility of a German plan to build a bomb and that he was indirectly responsible for the United States' own research into the matter. The report, written at the end of June 1945, summarized that a bomb had been built and that scientists and the military were expecting from day to day that the bomb would be exploded—that "a weapon has been developed that is potentially destructive beyond the wildest nightmares of the imagination." A brighter note, the report said, was that peaceful uses of atomic energy are possible as well. Most experts agree that the bomb would have been built by the United States even without Einstein's warning, simply because the technol-

Einstein in 1945

ogy had become available in the United States and in England.

In June, fifty nations signed the United Nations charter (Poland signed later), to go into effect October 24, and the League of Nations turned over its assets to the new organization.

Einstein officially retired from the Institute of Advanced Study and became a pensioner. The institute allowed him to keep his office in Fuld Hall until his death.

At a December 10 dinner at the Astor Hotel in New York honoring Nobel Prize winners, Einstein voiced his concerns about the bomb that physicists had delivered to the world that year. He soberly uttered the words that became world-famous: "The war is won, but the peace is not."

THIS YEAR THE NOBEL PRIZE in physics was awarded to Austria's Wolfgang Pauli for the discovery of the exclusion principle, also known as the Pauli principle. Artturi Vintanen of Finland received the award in chemistry for his research and inventions in agricultural and nutritional chemistry, especially for his fodder preservation method.

---

239. "Einstein on the Atomic Bomb." Edited by Raymond Swing. *Atlantic Monthly* 176 (November 1945): 43–45. Articles with the same title had appeared in the *New York Times* on October 27 and October 29. Reprinted as "Atomic War or Peace, Part 1," in *Ideas and Opinions* (1954), 118–123. See paper 251 for part 2.

Einstein advocated a world government for the purpose of controlling use of the atomic bomb and all armaments. It should be founded, he wrote, by the three great military powers: the United States, Soviet Union, and Great Britain. This world government would have power over all military matters and be able to intercede in countries where oppression occurs, because a world government is preferable to the far greater evil of wars. Later, the Russian Academy of Science denounced Einstein for advocating such a system (see *Ideas and Opinions*, 134–140).

The Emergency Committee of Atomic Scientists distributed a reprint of this article in November 1947 and January 1948 with an appeal for financial support.

---

240. "On the Cosmological Problem." *American Scholar* 14 (1945): 137–156. Corrections to the article are on p. 269 of the journal.

241. "Generalization of the Relativistic Theory of Gravitation." *Annals of Mathematics* 46, ser. 2 (1945): 578–584.

242. "Influence of the Expansion of Space on the Gravitation Fields Surrounding Individual Stars" (with E. G. Straus). *Review of Modern Physics* 17 (1945): 120–124; corrections and additions, 18 (1945): 148–149.

Ernst Straus was Einstein's collaborator at the Institute for Advanced Study. His recollections of working with Einstein can be found in A. P. French, ed., *Einstein: A Centenary Volume* (Cambridge: Harvard University Press, 1979).

243. "A Testimonial from Prof. Einstein." In Jacques Hadamard, *An Essay on the Psychology of Invention in the Mathematical Mind*, 142–143. Princeton, N.J.: Princeton University Press, 1945. Princeton Science Library edition: Appendix, in *The Mathematician's Mind: The Psychology of Invention in the Mathematical Field*. Princeton, N.J.: Princeton University Press, 1996.

Einstein answered a questionnaire that Hadamard, a French mathematician, had sent to mathematicians for a survey on how their mental processes work.

244. "On Harnessing Solar Energy." *New York Times*, August 14, 1945.

1946

In the postwar years, Einstein became increasingly disturbed about the atomic bomb and its use. Returning to his prewar pacifism, he spoke out in favor of a world government and sometimes advocated "Gandhi's methods" to achieve it. He believed that, as long as individual nations maintained separate armaments, there would always be wars. He took on the chairmanship of the newly formed Emergency Committee of Atomic Scientists, which was concerned with the peaceful uses of atomic energy. From now until the end of his life, he immersed himself in questions of peace and world government. He also supported the state of Israel, even though he did not always agree with its policies. As to his relationship with Germany, he would not reconcile with the country and wanted nothing to do with Germans except for the few who had been steadfast in their opposition to Hitler.

Einstein's sister, Maja, planned to return to Geneva after the war to rejoin her ill husband, but it was not to be. She suffered a stroke and remained bedridden, and the condition worsened over the years. Einstein faithfully read to his sister, who had come to resemble him greatly in old age, every night.

THIS WAS A YEAR of several achievements in physics. Lev Landau of the Soviet Union postulated an attenuation of wave motion when the velocity of a wave is comparable to the velocity of electrons in plasmas; he would win the Nobel Prize in 1962. Willard F. Libby began to develop radioactive carbon-14 dating, for which he received the Nobel Prize in chemistry in 1960. George Gamow suggested that the relative abundances of the elements had been determined by nucleosynthesis during the early stages of the universe's expansion. Fred Hoyle suggested that collapsing stars will continue to collapse until they become rotationally unstable and throw off the heavy elements they have built up. And John von Neumann and colleagues defined the concept of a software program and showed how a computer could execute such a program. Scientists at the University of Pennsylvania developed the ENIAC computer, containing eighteen thousand vacuum tubes, which demonstrated that high-speed digital computing was possible with vacuum-tube technology and laid the foundation for the electronic computer industry.

Einstein surveys the land.

The Nobel Prize in physics was awarded to Percy W. Bridgman of the United States for the invention of an apparatus to produce extremely high pressure and for the discoveries he made with it in high-pressure physics. One-half of the chemistry prize was awarded to James B. Sumner of the United States for his discovery that enzymes can be crystallized, and the other half was awarded jointly to John H. Northrup and Wendell M. Stanley of the United States for their preparation of enzymes and virus proteins in pure form.

245. "Social Obligation of the Scientist." In *Treasury for the Free World*, edited by R. Raeburn. New York: Arco, 1946.

This book originated from the files of *Free World*, a publication that disseminated the ideas of international leaders on the urgent problems of the time. While it was being prepared, the atomic bombs fell on Japan. In the book, sixty-one world-renowned figures, including Charles de Gaulle, Fiorello La Guardia, Marshal Tito, Julian Huxley, and Darryl Zanuck, gave their opinions on the current world situation. Einstein's essay is presented in the form of a questionnaire. He is asked (1) how scientists can make their influence felt toward the aim of world cooperation (through the establishment of an international body that presides over military matters); (2) if scientists should concern themselves with political matters (every citizen should express his convictions); (3) if there is a relationship between the progress of physics and mathematics and the progress of society (yes, because they bring about new technology and because they are efficient counterweights against a materialistic attitude); and (4) what can be done to undo the effects of Nazism ("the Germans can be killed or constrained, but they cannot be reeducated to a democratic way of thinking and acting within a foreseeable period of time").

246. "To the Jewish Students" (An die jüdischen Studenten). *Aufbau* 12, no. 1 (1946): 16.

247. "Generalization of the Relativistic Theory of Gravitation." *Annals of Mathematics* 47 (1946): 731–741.

248. "Elementary Derivation of the Equivalence of Mass and Energy." *Technion Journal* 5 (1946): 16–17.

249. "$E = mc^2$: The Most Urgent Problem of Our Time." *Science Illustrated* 1 (April 1946): 16–17. Reprinted in *Ideas and Opinions* (1954), 337.

An explanation of the formula for the equivalence of mass and energy in Einstein's own words for the general reader.

1947

Einstein continued his intensive commitment to arms control and the formation of a world government. His son Hans Albert was appointed professor of engineering at the University of California at Berkeley.

The British proposal to divide Palestine was rejected by both Arabs and Jews, so the question was turned over to the United Nations, which announced a plan for partition.

"The Scroll of the Rule," one of the better-preserved Dead Sea Scrolls

FIGHTER ACE and test pilot Chuck Yeager achieved the first supersonic speeds in the rocket-powered Bell-X1. Meanwhile, John Bardeen, Walter Brattain, and William Shockley invented the point-contact transistor at Bell Labs in New Jersey. Britain's Cecil Powell and his collaborators discovered the pion, predicted by Hideki Yukawa in 1935 as the mediator of a short-range nuclear force, opening the door to modern particle physics. Close on the heels of this discovery, a series of cosmic-ray experiments by George Rochester and Clifford Butler indicated that a new type of unstable, neutral elementary particle existed, the V-particle, now known as the K meson; their discovery can be seen as the first step toward understanding the quark nature of matter. Finally, "flying saucers" were reported to have crashed near Roswell, New Mexico.

The Nobel Prize in physics was awarded to Sir Edward V. Appleton of England for his investigations of the physics of the upper atmosphere, especially for his discovery of the so-called Appleton layer. The chemistry prize was also awarded to an Englishman, Sir Robert Robinson, for his investigations on plant products of biological importance, especially alkaloids.

Legendary pilot Chuck Yeager breaks the sound barrier in the Bell X-1.

250. "The Military Mentality." *American Scholar* 16 (1947): 353–354. Reprinted in *Ideas and Opinions* (1954), 132–134.

Einstein accused America of placing the importance of naked power above all other factors that affect relations among nations, resulting in a military mentality in government. He warned that Germany's similar attitude beginning with Bismarck and Kaiser Wilhelm II resulted in Germany's decline in less than a hundred years. Now, the military mentality is even more dangerous because of the more powerful weapons.

251. "Atomic War or Peace." As told to Raymond Swing. *Atlantic Monthly* 180 (November 1947): 29–33. Reprinted as "Atomic War or Peace, Part 2," in *Ideas and Opinions* (1954), 123–131. See paper 239 above for part 1.

The Emergency Committee of Atomic Scientists distributed a reprint of this article separately with a plea for financial support. Here Einstein spoke in favor of outlawing the atomic bomb and again advocated a world government for the purpose of averting wars.

252. [Response to the editor on Walter White's article, "Why I Remain a Negro."] *Saturday Review of Literature* 30 (November 1, 1947): 21.

253. [Open letter to the United Nations General Assembly on how to "break the vicious circle."] *United Nations World* 1 (October 1947): 13–14.

254. "The Problem of Space, Ether and the Field in Physics." In *Man and the Universe*, ed. Saxe Commins and R. N. Linscott. Vol. 4 of The Great Thinkers series. New York: Random House, 1947. Originally in Einstein's *The World as I See It* (1934), pp. 82–100, and reprinted in *Ideas and Opinions* (1954), 276–285.

Einstein discussed scientific concepts as they have historically related to space, giving mathematical examples.

1948

After many years of illness and hardship, Einstein's first wife, Mileva, died at the age of seventy-two in Zurich. This left Eduard, who was often confined to a mental institution because of his battle with schizophrenia, without the attention and love his mother had provided over the years. Family friends visited him, but he never heard from his father directly.

Einstein's severe stomach problems were continuing to aggravate him. When he was hospitalized at the end of the year with severe pains, his doctor discovered that he had a large aneurysm of the abdominal aorta, an abnormal dilation the size of a grapefruit. Surgery was not indicated, however.

Einstein, looking older and troubled

In January in India, when the peace-loving and serene seventy-nine-year-old Mahatma Gandhi, one of Einstein's great heroes, was on his way to evening prayer, he was assassinated by a Hindu extremist who considered him a dangerous apostle of nonviolence and of unity between Hindus and Muslims.

In the Near East, the state of Israel was finally established, with Chaim Weizmann as president and David Ben-Gurion as premier. In the United States, the Marshall Plan was instituted to aid European economic recovery.

ON JUNE 21, the modern stored-program computer was born when the University of Manchester's Small-Scale Experimental Machine, nicknamed "Baby," successfully executed its first program. It was the first computer to store a changeable user program in electronic memory and process it at electronic speed.

Sin-itiro Tomonaga, Victor Weisskopf, Julian Schwinger, and Richard Feynman independently invented different methods of making the renormalization calculations of quantum electrodynamics (QED) precise. Maria Goeppert-Mayer and, independently, Hans Jensen proposed the shell structure of the nucleus, in which nucleons are assumed to occupy shells analogous to electron shells. The long-playing (LP) record was invented, and the 200-inch refracting telescope was dedicated at Mount Palomar, north of San Diego, even though it did not become fully operational until 1949. Holography was invented when Dennis Gabor proposed a way of displaying a three-dimensional image of an object by splitting a light beam.

The Nobel Prize in physics was awarded to Britain's Lord Patrick M. S. Blackett for his development of the Wilson cloud chamber method and for the discoveries he made with it in the fields of nuclear physics and cosmic radiation. Arne W. K. Tiselius of Sweden received the chemistry prize for his research on electrophoresis and adsorption analysis, especially for his discoveries concerning the complex nature of the serum proteins.

255. [Letter on universal military training addressed to the chairman of the Senate committee.] U.S. Congress, Senate, Committee on Armed Services, Hearings on Universal Military Training, p. 257. Read March 24, 1948.

256. "Religion and Science: Irreconcilable?" *Christian Register* 127 (June 1948): 19–20. Reprinted in *Ideas and Opinions* (1954).

In this response to a greeting sent by the Liberal Ministers' Club of New York City, Einstein replied that the answer to the question is complicated because, though people can agree on what science is, they often differ on a definition of religion. The mythical aspect of religion—which is only one aspect of it—is the part that is most likely to cause conflict, and myths are not necessary for the pursuit of religious aims.

257. "On Receiving the One World Award." An address given at Carnegie Hall, April 27, 1948. Reprinted in *Ideas and Opinions* (1954), 146–147.

Einstein, distraught at the consequences of the war and the specter of continuous rearmament, proposed that there is only one path to peace and security: a supranational organization.

258. "Looking Ahead." *Rotarian*, June 1948, 8–10.
This article includes a reprinting of two articles in the *Bulletin of Atomic Scientists* on international understanding (4 [1948]: 1) and a reply to Soviet scientists (4 [1948]: 35–37) and statements on world government.

259. "Quantum Mechanics and Reality" (Quantenmechanik und Wirklichkeit). *Dialectica* 2 (1948): 320–324.

260. "Generalized Theory of Gravitation." *Reviews of Modern Physics* 20 (1948): 35–39. See paper 270 for a similar article.

261. "Atomic Science Reading List." In '*48: Magazine of the Year* (January), 60–61.
Stating that "it is not enough to know all about isotopes and pitchblende and plutonium" to understand atomic energy and its capabilities, Einstein selected and described six periodicals and books that cover scientific and historical sources as well as the "problems of peace, security, and the continued life of man with man." These are the *Bulletin of the Atomic Scientists*, a monthly publication, which he said was the most valuable single source of up-to-date information on atomic energy; Selig Hecht's *Explaining the Atom* (1947), an account of the scientific steps that led to our present knowledge on nuclear fission; John Hersey's *Hiroshima* (1946), a novel describing the effect of the atomic bomb on everyday people; Cord Meyer Jr.'s *Peace or Anarchy* (1947), a review of the complex problems connected with peace and security; Emery Reves's *The Anatomy of Peace* (1945), which discusses the issues of peace and the need for a world government to secure it; and Raymond Swing's *In the Name of Sanity*, which contains discussions of the events that followed the news that atomic energy had been discovered and is based on a series of broadcasts

1949

Early this year, after Einstein was discharged from a month's stay in the hospital, he spent a few weeks convalescing in Florida with his old friend, Dr. Walter Bucky, whom he had visited many times at his residence on Seventy-sixth Street in New York. He also took time to finish writing his "Autobiographical Notes" for inclusion in a volume edited by Paul Schilpp (see paper 264 below).

During the year of Einstein's seventieth birthday, three hundred scientists assembled in Princeton to pay tribute to him and hold a symposium on his contributions. When Einstein entered the auditorium, a respect-

Einstein turns seventy and the Institute for Advanced Study holds a celebration.

ful silence filled the hall, followed by an impassioned standing ovation.

Einstein continued to refuse the efforts of the Germans to reconcile with him by declining a foreign membership in the Max Planck Institute and the honorary citizenship of Ulm, the city of his birth. He would do the same in the case of West Berlin in 1952 and with the German section of the International Organization of Opponents of Military Service in 1953.

The United Nations admitted Israel into its ranks, and the new nation moved its capital from Tel Aviv to Jerusalem. Meanwhile, the USSR tested its first atomic bomb. In the United States, the cornerstone for the permanent headquarters of the United Nations was laid in New York City for the structure that would be completed in 1952.

THE 200-INCH-MIRROR Hale telescope at Mount Palomar, in San Diego County in California, dedicated the previous year, was put into operation. Edwin Hubble took the first photographic exposure in January, and in October the telescope was made available to astronomers from Caltech and the Carnegie Institution of Washington, D.C.

Freeman Dyson, in several articles, unified Richard Feynman's and Julian Schwinger's radiation theories.

Kurt Gödel reported his discovery of solutions for the field equations of general relativity that described "rotating universes" in which it is possible to travel into the past. Gödel, a famous logician and eccentric who had escaped the Nazis and came to the Institute for Advanced Study in 1939, became Einstein's collaborator and good friend. He was best known for his incompleteness theorem of 1931, which implied that all complex logical systems are, by definition, incomplete; at any one time, each system contains more true statements than it can prove according to its own defining set of rules. They also lead to statements such as "this theorem is not a theorem." Over the years, Einstein and Gödel could be seen walking together along the streets of Princeton, Einstein happily smoking his pipe and the slight, bespectacled Gödel solemnly staring down at the ground, both deep in conversation. In his later years, Gödel apparently suffered from what is today known as obsessive-compulsive disorder, being paranoid about the spread of germs, compulsively cleaning his silverware before eating, and wearing a ski mask wherever he went. He suffered two nervous breakdowns, and, in the end, he refused to eat and died at age seventy-two in Princeton.

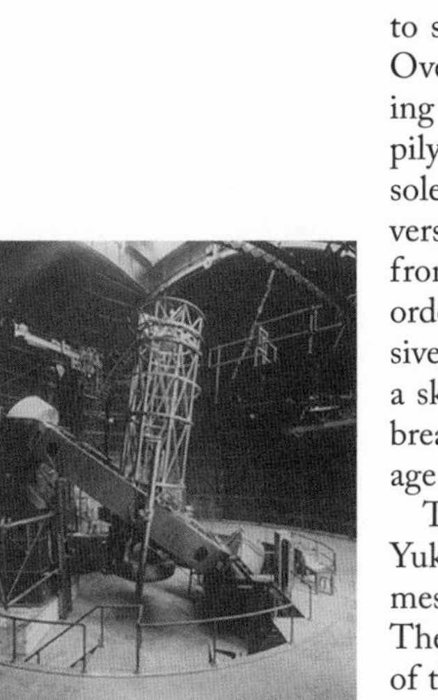
Mount Wilson Observatory

The Nobel Prize in physics was awarded to Hideki Yukawa of Japan for his prediction of the existence of mesons based on his theoretical work on nuclear forces. The chemistry prize was awarded to William F. Giauque of the United States for his contributions in the field of chemical thermodynamics, particularly concerning the behavior of substances at extremely low temperatures.

262. "Why Socialism?" *Monthly Review: An Independent Socialist Magazine* 1, no. 1 (May 1949): 9–15. Reprinted in *Ideas and Opinions* (1954), 151–158.

In this essay, Einstein expressed the view that the "economic anarchy" of capitalist society is a source of evil because individuals depend on society (which he called "a fact of nature") and that individuals are crippled (i.e., they can lose their jobs) by the profit motive and the competition of capitalists. He advocated the establishment of a socialist economy, where the means of production are owned by society in a planned way.

263. "In the Shadow of the Atomic Bomb." *Southern Patriot* 7, no. 3 (May 1949).

264. "Autobiographical Notes" and "Remarks to the Essays Appearing in This Volume." In *Albert Einstein: Philosopher-Scientist*, edited by Paul A. Schilpp, pp. 3–94 (in German and English) and 665–688. Library of Living Philosophers, vol. 7. La Salle, Ill.: Open Court, 1949.

Einstein referred to these notes, which he wrote in 1946 and which are essentially a short scientific biography, as his "obituary."

265. "Foreword." In *The Hebrew University of Jerusalem, 1925–1950*. Jerusalem: Goldberg's Press.

266. "On the Motion of Particles in General Relativity Theory" (with Leopold Infeld). *Canadian Journal of Mathematics* 3 (1949): 209–214.

The authors show that the field equations alone suffice as a basis for general relativity theory.

1950

Einstein, becoming ever more cognizant of his ill health and age, signed and sealed his last will, naming his friend Otto Nathan as executor and, along with Helen Dukas, trustee of his literary estate. He stipulated that his papers be donated to the Hebrew University of Jerusalem after the death of Nathan and Dukas, though arrangements were made later for an earlier transfer. He willed his violin to his grandson, Bernhard. His monetary assets would go to Helen Dukas; his sons, Hans Albert and Eduard; and his stepdaughter, Margot.

In February, President Truman announced that the United States had successfully built a hydrogen bomb—a bomb even more powerful than the atomic bomb. Hungarian-born physicist Edward Teller was dubbed its "father." Upon hearing the news, Einstein decided to speak in a nationwide television broadcast. He warned that if the bomb were ever used, the planet would be poisoned by radioactivity and all life would be annihilated.

In the United States, Senator Joseph McCarthy began his Communist witch-hunting by claiming that the State Department was infiltrated with Communists and

Einstein and J. Robert Oppenheimer

Communist sympathizers, though none were found. Over the next four years, he continued to look for and demand names of people from a variety of occupations and backgrounds, particularly entertainers and intellectuals, who he believed were somehow affiliated with the Communist Party. These people were required to testify in Congress and defend themselves at meetings of the House Un-American Activities Committee (HUAC). Einstein advised many of them not to cooperate, as he viewed this process to be a violation of their civil rights.

FRED HOYLE COINED the term *big bang* for the new theory about the creation of the cosmos. Even though he popularized the theory by giving it a name, he challenged the belief that the cosmos began with a primal fireball 12 million years ago. John Nash, who has become a popular icon since the book and film *A Beautiful Mind* appeared, introduced the concept of "noncooperative games" in game theory.

Germany had won the greatest number of Nobel Prizes by this time, with the United States in second place. This year the Nobel Prize in physics went to an Englishman, Cecil Powell, for his development of the photographic method of studying nuclear processes and for his work on the detection of the elementary particles known as mesons. The chemistry prize went to two Germans, Otto Diels and Kurt Adler, for their discovery and development of diene synthesis.

267. "On the Moral Obligation of the Scientist." *Impact* 1 (1950): 104–105.

268. "Physics, Philosophy, and Scientific Progress." Tape-recorded address to a convention of the International College of Surgeons. Also in *Journal of the International College of Surgeons* 14 (1950): 755–758.

269. An Open Letter to the Society for Social Responsibility in Science. *Science* 112 (1950): 760–761. Reprinted in *Ideas and Opinions* (1954) as "The State and the Individual Conscience," 26–27.

In this letter of December 22, 1950, Einstein praised the formation of the society and reminded the members that "institutions are in a moral sense impotent unless they are supported by the sense of responsibility of living individuals." The Nuremberg trials had made it self-evident that individual responsibility cannot be expunged.

270. "On the Generalized Theory of Gravitation." *Scientific American* 182, no. 4 (April 1950): 13–17. Reprinted in *Ideas and Opinions* (1954).

At the request of the editors of the magazine, Einstein discussed his mathematical investigations into the foundations of field physics.

271. "The Bianchi Identities in the Generalized Theory of Gravitation." *Canadian Journal of Mathematics* 4 (1950): 120–128.

272. "Foreword." In Philipp Frank, *Relativity: A Richer Truth*. London: Jonathan Cape, 1951.

In the foreword to this small book on the ethical implications of relativity, Einstein wrote that it may seem that logical thinking is not relevant for ethics but that, indeed, "ethical directives can be made rational and coherent by logical thinking and empirical knowledge." Ethical axioms, like the axioms of science, are founded and then tested; if they fit society or an individual, they are accepted.

273. "Message to the Italian Society for the Advancement of Science." Sent to the group's forty-second meeting in Lucca, Italy. Published in English in the UNESCO journal *Impact*, Fall 1950.

In this address, Einstein wrote about the search for truth and understanding, saying that such inquiry attempts to encompass the complex experiences of humankind while looking for simplicity and economy in basic assumptions. The man of science has suffered a tragic fate because, while seeking to strive for truth and understanding, he has also developed destructive and enslaving tools.

Senator Joseph McCarthy's "Red Scare" ruins the lives of many academics, artists, politicians, and labor leaders.

1951 1952

After her long confinement to bed, Maja died in Princeton in 1951. Einstein missed her enormously. Her husband, Paul, died the following year in Geneva.

In a letter to Queen Elisabeth of Belgium, Einstein mentioned that he was still working on a unified field theory and that he would continue to do so until he took his last breath. In his unified field theory, Einstein attempted to express gravitational theory and electromagnetic theory within a single unified framework, to show that the two forces are based on one grand underlying principle. He had trouble accepting two distinct forces in nature—and this was even before the discovery of two additional forces: the strong and weak nuclear interactions. Today, physicists are searching for an even loftier theory: a Theory of Everything (TOE), which they hope will unify all forces and all matter. At the moment, string theory appears to be the tool of choice. For a popular exposition of this research, see Brian Greene's *The Elegant Universe* (2001).

After Israel's first president, Chaim Weizmann, died in 1952, Einstein was offered the presidency of the new nation. He declined, saying that he was deeply honored by the gesture but not suited to the job. Israeli leaders, knowing that Einstein's skills lay elsewhere, were relieved that he turned down the offer. Prime Minister David Ben-Gurion pleaded with Yitzak Navon, who in 1978 became president of Israel himself: "Tell me what to do if he says yes. I had to offer the post to him because it is impossible not to. But if he accepts, we're in for trouble." Though he was not a Zionist, Einstein supported Israel while lamenting the fact that the nation was established through violence. He was also concerned about the prospect for the long-term peaceful coexistence between Jews and Arabs.

Israel and Germany agreed on reparations for damages inflicted by the Nazis on European Jews.

IN 1951, Rosalind Franklin discovered the nucleic acids RNA and DNA and elucidated the basic helical structure of the DNA molecule. Though she was the first to make these discoveries, her data were given to and used by others, and the Nobel Prize for the discovery was awarded

Einstein with Kurt Gödel in Princeton

Bohr and Wolfgang Pauli play with a new toy, the "tippie top."

to three scientists in 1962 for the ultimately correct and detailed description of DNA's structure, four years after her death in 1958 of ovarian cancer. Nobel Prizes are not awarded posthumously.

As in 1919, another total eclipse of the Sun took place in February 1952, visible from Europe, enabling astronomers again to confirm the effect of gravity on light as predicted by relativity theory. On November 18, 1952, the United States tested its first hydrogen bomb on Enewetok Atoll in the Marshall Islands in the Pacific Ocean.

The Nobel Prize in physics for 1951 was won jointly by Sir John Cockcroft of England and Ernest T. S. Walton of Ireland for their pioneering work on the transmutation of atomic nuclei by artificially accelerated atomic particles. The chemistry prize went to Americans Edwin McMillan and Glenn Seaborg for their discoveries in the chemistry of transuranium elements.

The Nobel Prize in physics for 1952 was awarded jointly to Americans Felix Bloch and Edward Purcell for their development of new methods for nuclear magnetic precision measurements and discoveries in connection therewith. The chemistry prize was awarded jointly to Britain's Archer J. P. Martin and Richard Laurence for their invention of partition chromatography.

274. "Foreword." In Dagobert D. Runes, ed., *Spinoza: Dictionary.* New York: Philosophical Library, 1951.

Einstein praised this dictionary of Spinoza's terminology, given in Spinoza's own words, as a reliable guide to Spinoza's works, which are otherwise often difficult to understand. Baruch Spinoza, a seventeenth-century Jewish theologian and philosopher living in Holland, had greatly influenced Einstein's religious thinking. Spinoza dealt negatively with orthodox Jewry, preaching that the laws of the Torah were obsolete, for which he was excommunicated from the orthodox ranks.

275. "The Advent of the Quantum Theory." *Science* 113 (1951): 82–84.

276. "Introduction." In Carola Baumgardt, *Johannes Kepler: Life and Letters.* New York: Philosophical Library, 1951.

In this collection of annotated letters from Kepler to his contemporaries, including Galileo, Einstein wrote an introduction tracing Kepler's determination of the movement of Earth in planetary space. Because Earth "itself can be used at any time as a triangulation point, Kepler was also able to determine by observation the true movements of the other planets." The letters show that Kepler completed his work under conditions of great personal hardship.

277. "Foreword." In Homer Smith, *Man and His Gods.* Boston: Little, Brown, 1952.

In this history of religious and philosophical thinking, or a person's "place in nature and the universe," by a prominent biologist, Einstein praised Smith's objectivity in presenting his subject, saying that such objectivity is rarely found in a "pure historian." Smith does not neglect to recount the suffering that "man's mythic thought" has often wrought on humankind.

278. "Those Who Read Only Newspapers See Things Like a Nearsighted Person without Glasses" (Wer nur Zeitungen liest, sieht die Dinge wie ein Kurzsichtiger ohne Augengläser). *Der Jungkaufmann* (Zurich) 27, no. 4 (1952): 73. See Einstein's *Ideas and Opinions* (1954), 64–65, for an English translation.

In this letter, Einstein stated his belief that it is important for young people to read the classics. Because there are only a few exceptional people of letters in any given century, their work becomes "the most precious possession of mankind."

279. "Symptoms of Cultural Decay." *Bulletin of Atomic Scientists* 8, no. 7 (October 1952). Reprinted in *Ideas and Opinions* (1954), 166–167.

According to Einstein, politicians should neither control the sciences nor impede free scientific exchange with other countries. Politicians are now so distrustful, he wrote, that even in peacetime they are determined to organize our lives and work as if we are preparing to win a war.

1953

This year was not eventful in Einstein's personal life. He became less physically active as his health deteriorated, and he was seldom seen walking the streets of downtown Princeton, eating ice cream cones, petting dogs, and chatting with neighbors.

IN THE HIMALAYAS, New Zealander Edmund Hillary and Nepalese mountain guide Tenzing Norgay became the first people to reach the 29,028-foot summit of Mount Everest, a testament to human endurance and cooperation. In the sciences, Russia's Andrei Sakharov invented a fusion and fission detonator, the basis for the first thermonuclear bomb built by the Soviet Union. His work on this superbomb was independent of the work of America's Edward Teller and Stanislaw Ulam. Murray Gell-Mann and, independently, Kazuhiko Nishijima introduced the concept of "strangeness," a quantum property that accounted for previously puzzling decay patterns in some mesons. In 1961 Gell-Mann would develop the eightfold-way theory, and in 1969 he would receive the Nobel Prize in physics.

The Nobel Prize in physics was awarded to Fritz

James Watson and Francis Crick discover the structure of DNA.

Tenzing Norgay and Edmund Hillary conquer Mount Everest.

Zernike of the Netherlands for his demonstration of the phase contrast method, especially for his invention of the phase contrast microscope. The chemistry prize went to Germany's Hermann Staudinger for his discoveries in the field of macromolecular chemistry.

280. "To the Jewish Peace Fellowship." Dated September 21, 1953. *Tidings* 8, no. 1 (1953): 3. Also in *Tidings* 9, no. 2 (1955): 5.

281. "A Comment on a Criticism of a Recent Unified Field Theory." *Physical Review* 89 (1953): 321.

This was Einstein's final paper on the unified field theory. He was certain about the mathematical concepts but uncertain about the physical aspects.

282. "Elementary Reflections on the Interpretation of the Foundations of Quantum Mechanics" (Elementare Überlegungen zur Interpretation der Grundlage der Quantenmechanik). In *Scientific Papers Presented to Max Born.* Edinburgh: Oliver and Boyd, 1953.

283. "Preface." In Galileo Galilei, *Dialogue Concerning the Two Chief World Systems: Ptolemaic and Copernican*. Translated by Stillman Drake. Berkeley and Los Angeles: University of California Press, 1953.

284. "Letter in Reply to William Frauenglass." *Bulletin of the Atomic Scientists* 9 (1953): 230. Reprinted in *Ideas and Opinions* (1954), 33–34.

In this letter of May 16, 1953, Einstein encouraged Brooklyn teacher William Frauenglass to refuse to submit to questioning by the U.S. Senate's Internal Security Subcommittee. All intellectuals should refuse to testify, not on the basis of the Fifth Amendment but because it is shameful for an innocent citizen to have to submit to such an inquisition, a violation of the spirit of the Constitution.

285. "Message to the 24th Annual Conference of the War Resister's League," New York, August 10, 1953.

1954

McCarthyism gets out of hand, and Senator McCarthy is officially censured by the U.S. Senate.

Einstein, though he developed hemolytic anemia, worked on his last scientific paper with collaborator Bruria Kaufmann, publishing it the following year (paper 293).

At the height of McCarthy's House Un-American Activities Committee (HUAC) hearings and the Oppenheimer case, the U.S. government alleged that former Los Alamos director J. Robert Oppenheimer had Communist sympathies and therefore withdrew his security clearance, ending his influence on scientific policy. Einstein supported Oppenheimer and continued to encourage others to refuse to testify in front of the investigating committees, becoming a hero of the left and liberals and an antihero of the right, which wanted him stripped of his citizenship and deported. The specter of intellectuals being persecuted and civil liberties and freedom of expression being denied was painful for the aging scientist, who was all too familiar with this situation.

The witch-hunting by Senator Joseph McCarthy continued in televised hearings. McCarthy now wanted to prove Communists had infiltrated the U.S. Army. The attorney for the army famously asked him, "Have you no shame?" McCarthy was finally censured and condemned by a Senate resolution for his insults to the Senate. Since then, McCarthyism has become synonymous with unfounded political accusations and witch-hunting with unsupported evidence.

THE UNITED STATES tested another hydrogen bomb, now on Bikini Atoll in the Marshall Islands, vaporizing three of the islands. Oppenheimer—and Einstein—had opposed this project. Americans and Europeans—not to mention the Pacific Islanders—expressed concern about radioactive fallout from nuclear testing.

In the sciences, Jonas Salk of the United States developed a vaccine against polio. After mass immunizations the following year, the incidence of polio began to decline dramatically.

Important discoveries were made in physics: Frank Yang and Richard Mills discovered non-Abelian gauge theory, which initiated a new way of thinking in particle physics; Robert Hofstadter began his pioneering studies of electron scattering in atomic nuclei, enabling researchers to find the size and surface-thickness parameters of nuclei (he would win the Nobel Prize in physics in 1961); and Charles Townes and collaborators demonstrated the maser principle, leading six years later to the first laser (Townes would be awarded the Nobel Prize in physics in 1964). Meanwhile, mathematician John von Neumann at the Institute for Advanced Study was developing a theory of artificial automata, which process information and act according to internal rules and instructions, just like their natural counterparts.

The Nobel Prize in physics was divided between Germany's Walther Bothe, for the coincidence method and his discoveries made therewith, and Einstein's old friend Max Born, now living in Scotland, for his fundamental research in quantum mechanics, especially for his sta-

Einstein makes a point.

tistical interpretation of wave function. The chemistry prize went to Linus Pauling of the United States for his research into the nature of the chemical bond and its application to elucidation of the structure of complex substances.

286. *Ideas and Opinions*. Translated by Sonja Bargmann. New York: Crown, 1954.
This is the most popular compilation of a variety of Einstein's writings, still in print.

287. "Algebraic Properties of the Field in the Relativistic Theory of the Asymmetric Field" (with Bruria Kaufmann). *Annals of Mathematics* 59 (1954): 230–244.

288. "Foreword." In Max Jammer, *Concepts of Space*. Cambridge: Harvard University Press, 1954.

289. *The Meaning of Relativity*. 5th ed. Princeton: Princeton University Press, 1954.
This is the final revised edition of Einstein's seminal work. In it, he completely revised appendix 2, "Generalization of Gravitation Theory," of the fourth edition, renaming it "Relativistic Theory of the Non-symmetric Field"—his final attempt to extend or generalize his theory to achieve a unified theory of the gravitational and electromagnetic fields.

290. "If Einstein Were Young Again, He Says He'd Become a Plumber." *New York Times*, November 10, 1954.

This is a report about a letter Einstein wrote to the editor of *The Reporter* magazine on October 13, in which Einstein claimed that if he could live his life again, he would not become a scientist. He said he would rather become a plumber or a peddler, "in the hope of finding that modest degree of independence still available under present circumstances." Unconstrained independence would allow him to concentrate freely on physics.

291. "Tribute to Joseph Scharl." *New York Times*, December 9, 1954.
Einstein praised the artist who had sketched him.

1955

Early in the year, Bertrand Russell approached Einstein, asking him to issue a joint statement with other scientists on the arms race, declaring that in a nuclear war there would be no winners or losers, only a permanent state of catastrophe. Russell drafted the statement and sent it to Einstein, who signed it on April 11 and returned it to Russell. With it, he sent a short letter.

A week later, early in the morning, Einstein died at Princeton Hospital after suffering a ruptured abdominal aneurysm. Alberta Rozsel, the night nurse, was the last person to see him alive. His body, from which the brain and eyes had been removed during the autopsy, was cremated late in the afternoon, and his ashes were scattered, probably over the Delaware River not far from Princeton, by his friends Otto Nathan and Paul Oppenheim. The news spread rapidly throughout the world as a flood of tributes filled the media.

The restaurant that started it all, the original McDonald's

IN PHYSICS, Emilio Segrè discovered "anti-protons" (he would win the Nobel Prize in 1959), and John Wheeler described a hypothetical object, a "geon," constructed out of electromagnetic radiation.

The Nobel Prize in physics was divided equally between Americans Willis E. Lamb, for his discoveries concerning the fine structure of the hydrogen spectrum, and Polykarp Kusch for his precise determination of the magnetic moment of the electron. The chemistry prize went to Vincent du Vigneaud, also of the United States, for his work on biochemically important sulfur compounds, especially for the first synthesis of a polypeptide hormone.

292. "Foreword." In Louis de Broglie, *Physics and Microphysics*. New York: Pantheon, 1955.

293. "Preface." In Jules Moch, *Human Folly: To Disarm or Perish*. London: Gollancz, 1955.

294. "A New Form of the General Relativistic Field Equations" (with Bruria Kaufmann). *Annals of Mathematics* 62 (1955): 128–138.

295. "The Einstein-Russell Manifesto." Issued from London, July 9, 1955. Also signed by Max Born, Leopold Infeld, Frédéric Joliot-Curie, and Linus Pauling, among others, the manifesto can be seen on the Internet at www.nuclearfiles.org.

296. "Remembrances" (Erinnerungen-Souvenirs). *Schweizerische Hochschulzeitung* 28 (special volume, 1955): 143–153.

Einstein recalled, among other things, that he had studied the work and lives of the "masters of theoretical physics" while a student.

297. "Albert Schweitzer at Eighty." *Christian Century* 72, no. 2 (1955): 42.

298. "Introduction." In Henry I. Wachtel, *Security for All and Free Enterprise*. New York: Philosophical Library, 1955.

299. "Foreword." In William Esslinger, *Politics and Science*. New York: Philosophical Library, 1955.

300. "What Is the Task of an International Journal?" *Common Cause* (Florence, Italy) 1, no. 1: 3.

BIBLIOGRAPHY

Relief carving of Einstein at the Riverside Church, New York City

For Einstein's early papers I relied heavily on scientific information gleaned from *The Collected Papers of Albert Einstein (CPAE),* volumes 1–4, 6, and 7, and their editors John Stachel, Martin Klein, Michel Janssen, Anne Kox, and Robert Schulmann and their editorial staff, and on *Einstein's Miraculous Year,* edited and translated by John Stachel (with the assistance of Trevor Lipscombe, Alice Calaprice, and Sam Elworthy), for the papers of 1905. Also useful was John Stachel's *Einstein from B to Z,* Sachi Sri Kantha's *An Einstein Dictionary,* and, at the last minute, the difficult-to-find book by Cornelius Lanczos, *The Einstein Decade (1905–1915).* Finally, *In Albert's Shadow: The Life and Letters of Mileva Marić, Einstein's First Wife,* edited by Milan Popović, was very helpful by providing insights into Albert's family life from Mileva's perspective.

For a balanced view of the hole argument (paper 40), see particularly John Stachel's paper, "Einstein's Search for General Covariance, 1912–1915," in Stachel and Howard, eds., *Einstein and the History of General Relativity* (Birkhäuser, 1989). Also see work by John Norton and John Earman.

English translations for the years 1901–21 can be found in the translation volumes of *CPAE,* also published by Princeton University Press and available at most university libraries. When I was able to find English translations for the years after 1921, I added the sources.

I consulted *CPAE* and several Einstein biographies, particularly Albrecht Fölsing's *Albert Einstein,* Dennis Overbye's *Einstein in Love,* and Abraham Pais's *Subtle Is the Lord,* for biographical material. I also drew from my twenty-five years' experience working with the Einstein Papers. Most of the works are listed in Nell Boni et al.'s *Readex Checklist,* which was a great help, as were the bibliographies in the Pais and Fölsing biographies. Many of the popular writings, both scientific and nonscientific, suitable for a general audience are in English in *Ideas and Opinions.* And, last but not least, I hope my *Expanded Quotable Einstein* or the forthcoming new edition, *The New Quotable Einstein* (2005), is useful to those who want a very concise look at Einstein's views on many topics.

For general overviews in physics, I can recommend Jeremy Bernstein's *Quantum Profiles,* Brian Greene's *Elegant Universe,* Sam Schweber's *QED and the Men Who Made It,* Julian Schwinger's *Einstein's Legacy,* Harriet Zuckerman's *Scientific Elite,* and Gary Zukav's *The Dancing Wu Li Masters.*

Also of enormous help with historical and scientific information were the Websites nature.com, nobel.se, nuclearfiles.org, atomicmuseum.com, cerncourier.com, aps.org, aip.org, crimsonbird.com, almaz.co, userpage.chemie.fu-berlin.de, lbl.gov, lns.cornell.edu, seop.leeds.ac.uk, and sciencetimeline.net.

Bernstein, Jeremy. *Quantum Profiles.* Princeton, N.J.: Princeton University Press, 1991.

Boni, Nell, Monique Russ, and Dan H. Laurance. *A Bibliographical Checklist and Index to the Collected Writings of Albert Einstein.* New York: Readex Microprint, 1960.

Bonner, Thomas N. *Iconoclast: Abraham Flexner and a Life in Learning.* Baltimore: Johns Hopkins University Press, 2002.

Calaprice, Alice. *The Expanded Quotable Einstein.* Princeton, N.J.: Princeton University Press, 2000.

Collected Papers of Albert Einstein (CPAE). Vols. 1–9. Princeton, N.J.: Princeton University Press, 1986–2004.

Einstein, Albert. *Ideas and Opinions.* New York: Crown, 1954.

———. *The World as I See It.* New York: Covici Friede, 1934.

Fantova, Johanna. "Conversations with Einstein." Manuscript in Department of Rare Books and Special Collections, Princeton University Library.

Fölsing, Albrecht. *Albert Einstein.* New York: Viking, 1997.

Gardner, Howard. *Creating Minds: An Anatomy of Creativity,* 87–131. New York: Basic Books, 1993.

Greene, Brian. *The Elegant Universe.* Also a PBS series. New York: Vintage Books, 2001; reprint, New York, Norton, 2003.

Jerome, Fred. *The Einstein File: J. Edgar Hoover's Secret War against the World's Most Famous Scientist.* New York: St. Martin's, 2002.

Kantha, Sachi Sri. *An Einstein Dictionary.* Westport, Conn.: Greenwood Press, 1996.

Lanczos, Cornelius. *The Einstein Decade (1905–1915).* London: Elek Science, 1974.

Nathan, Otto, and Heinz Norden, eds. *Einstein on Peace.* New York: Simon and Schuster, 1960.

Norton, John. "How Einstein Found His Field Equations, 1912–1915." In John Stachel and Don Howard, eds., *Einstein and the History of General Relativity,* 101–159. Einstein Studies, vol. 1. Boston: Birkhäuser, 1989.

Overbye, Dennis. *Einstein in Love: A Scientific Romance.* New York: Viking, 2000.

Oxford Dictionary of Physics. 4th ed. Edited by Alan Isaacs. Oxford: Oxford University Press, 2000.

Pais, Abraham. *Subtle Is the Lord . . . : The Science and the Life of Albert Einstein.* New York: Oxford University Press, 1982.

Popović, Milan. *In Albert's Shadow: The Life and Letters of Mileva Marić, Einstein's First Wife.* Baltimore: Johns Hopkins University Press, 2003.

Roboz-Einstein, Elizabeth. *Hans Albert Einstein.* Iowa City: University of Iowa, 1991.

Rosenkranz, Ze'ev. *The Einstein Scrapbook.* Baltimore: Johns Hopkins University Press, 2002.

Schilpp, Paul, ed. *Albert Einstein: Philosopher-Scientist.* Evanston, Ill.: Library of Living Philosophers, 1949.

Schweber, Silvan S. *QED and the Men Who Made It.* Princeton, N.J.: Princeton University Press, 1994.

Schwinger, Julian. *Einstein's Legacy.* New York: Scientific American Books, 1986.

Stachel, John. *Einstein from B to Z.* Boston: Birkhäuser, 2002.

Stachel, John, ed., with the assistance of Trevor Lipscombe, Alice Calaprice, and Sam Elworthy. *Einstein's Miraculous Year.* Princeton, N.J.: Princeton University Press, 1998.

Zuckerman, Harriet. *Scientific Elite.* New York: Free Press, 1977.

Zukav, Gary. *The Dancing Wu Li Masters: An Overview of the New Physics.* New York: Perennial Classics, 2001.

INDEX

Italic page numbers indicate illustrations.

ILLUSTRATION CREDITS

Page iii. Photo ca. 1950s. AIP Emilio Segrè Visual Archives, Landé Collection.
Page xi. *The Papers of Thomas A. Edison,* volume 5, *Research to Development at Menlo Park, January 1879–March 1881,* ed. Paul B. Israel, Louis Carlat, David Hochfelder, and Keith A. Nier (Baltimore: Johns Hopkins University Press, 2004), 211.
Page xii. Photo ca. 1865–80. Brady-Hardy Photograph Collection, Library of Congress (LOC).
Page xiii. Reprinted from an unidentified English newspaper, 1886. LOC Prints and Photographs Collection.
Page xiii. Handbill posted throughout Chicago and printed in *Arbeiter-Zeitung,* an anarchist newspaper published by August Spies, May 3, 1886. LOC Prints and Photographs Collection.
Page xv. From *Harper's Weekly,* sketches by G. A. Coffin and Charles Mente, 1894. LOC Prints and Photographs Collection.
Page xvi. "A View of Our Battleship *Maine* as She Appears To-day, May 10, 1900, Havana Harbor." LOC Prints and Photographs Collection.
Page xvii. "Restoration and Defense of British Liberty in South Africa," 1900. LOC Prints and Photographs Collection.
Page 1. "Europe at the Present Time," from *The Historical Atlas,* by William R. Shepherd, 1911.
Page 5. Photograph of Bern, no date. From the author's collection.
Page 5. Nobel Prize for Physics, 1921. © The Nobel Foundation.
Page 7. *In Albert's Shadow: The Life and Letters of Mileva Marić, Einstein's First Wife,* ed. Milan Popović (Baltimore: Johns Hopkins University Press, 2003), 40.
Page 8. *In Albert's Shadow: The Life and Letters of Mileva Marić, Einstein's First Wife,* ed. Milan Popović (Baltimore: Johns Hopkins University Press, 2003), 47.
Page 10. "First Flight, 120 feet in 12 seconds, 10:35 am; Kitty Hawk, North Carolina," December 17, 1903. Glass negatives from the papers of Wilbur and Orville Wright, LOC Prints and Photographs Collection.
Page 12. Einstein, Bern, Switzerland, 1905. Lotte Jacobi Collection, University of New Hampshire.
Page 14. "President Theodore Roosevelt passing 10th Street and Pennsylvania Avenue in inauguration parade on the way to the Capitol," 1905. National Photo Company, LOC Prints and Photographs Collection.
Page 17. "San Francisco Earthquake," a view of the aftermath on Market Street facing east, 1906. George R. Lawrence Company, LOC Prints and Photographs Collection.
Page 27. From "Halley's Comet: A Bibliography," compiled by R. Freitag, 1984. Original art by Fernand Baldet, 1910. LOC Prints and Photographs Collection.
Page 30. "Titanic," 1911. George Grantham Bain Collection, LOC.
Page 31. "Discovery and explorations of the South Pole by Capt. Roald Amundsen and crew," 1910–11, published May 23, 1912, by United Newspapers, Ltd., London. LOC Prints and Photographs Collection.
Page 33. "Insurrectos Outpost," Mexican Revolution, ca. 1911, published by American Press Association, March 9, 1911. LOC Prints and Photographs Collection.
Page 36. "Suffragette Parade, March 3rd 1913." George Grantham Bain Collection, LOC.
Page 38. "Panama Canal Worksite," October 15, 1913. LOC Prints and Photographs Collection.
Page 43. "Elizabeth Gurley Flynn at trial for inciting strikers to violence, Nov. 29, 1915." LOC Prints and Photographs Collection.
Page 44. "1915, Deported Armenian family—two older couples and two young children—living under a tent in the desert. Location: Ottoman Empire, region: Syria," by Armin T. Wegner. © Wallstein Verlag, Göttingen, Federal Republic of Germany. All rights reserved.
Page 47. *In Albert's Shadow: The Life and Letters of Mileva Marić, Einstein's First Wife,* ed. Milan Popović (Baltimore: Johns Hopkins University Press, 2003), 44.
Page 49. Vladimir Ilyich Lenin, ca. 1920. Soyuzfoto, LOC Prints and Photographs Collection.
Page 52. "Tsar Nicholas of Russia, his wife and their five children," ca. 1910–18. *New York World-Telegram* and the *Sun Newspaper* Photograph Collection, LOC.
Page 55. LOC Prints and Photographs Collection.
Page 56. "Boxer Rebellion: Boxers before the High Court, China," published December 13, 1919. LOC Prints and Photographs Collection.
Page 59. "Miss Alice Pavl is shown sewing the thirty-sixth star on the suffrage ratification banner, the stars having been added from time-to-time as the various states ratified," 1920. National Photo Company, LOC Prints and Photographs Collection.
Page 62. Brown Bros., Sterling, Pennsylvania, 1921.
Page 63. LOC Prints and Photographs Collection.
Page 64. Photo by Willem J. Luyten, 1921. American Institute of Physics.
Page 64. Nobel Prize for Physics, 1921. © The Nobel Foundation.
Page 68. Niels Bohr, 1922. Niels Bohr Archives, Copenhagen. All rights reserved.
Page 71. "Ships going through the Panama Canal, West Lirio side," 1923. Frank and Frances Carpenter Collection, LOC.
Page 72. "Prohibition officers raiding the Lunch Room of 922 Pa. Ave, Wash. D.C.," April 25, 1923. National Photo Company Collection, LOC.
Page 72. Sigmund Freud, 1856–1939 (photo 1938). LOC Prints and Photographs Collection.
Page 74. Marcus Garvey, August 5, 1924. George Grantham Bain Collection, LOC.
Page 75. Babe Ruth, 1924, Yankees versus Senators. National Photo Company Collection, LOC.
Page 76. J. Edgar Hoover, December 22, 1924. National Photo Company Collection, LOC.
Page 81. Marcel Louis Brillouin, Marie Curie, Albert Einstein, Paul Langevin, Hendrik Antoon Lorentz, Jean Baptiste Perrin, Max Karl Ernst Ludwig Planck, and Ernest Rutherford at the First Solvay Congress, Brussels, 1911. Photo by Benjamin Couprie, Institut International de Physique Solvay. AIP Emilio Segrè Visual Archives.
Page 86. Einstein Archives.
Page 87. AIP Emilio Segrè Visual Archives.

Page 89. Cartoon by C. Berryman, 1929. The *Washington Post* Writer's Group. © 1929, *The Washington Post.* Reprinted with permission.
Page 92. "Two iron workers straddle steel girders on top of the Empire State building as it nears completion," 1931. *New York World-Telegram* and the *Sun Newspaper* Photograph Collection, LOC.
Page 93. © *New York Times* Pictures.
Page 94. "Gandhi at his spinning wheel aboard ship en route to London," 1931. Associated Press photo, *New York World-Telegram* and the *Sun Newspaper* Photograph Collection, LOC.
Page 97. © AP/Wide World Photos.
Page 98. Archives Lemaître, Université catholique de Louvain, Institut d'Astronomie et de Géophysique G. Lemaître, Louvain-la-Neuve, Belgique.
Page 100. Adolph Hitler and Hitler Youth, Erfurt, Germany. LOC Prints and Photographs Collection.
Page 103. Paul Adrien Maurice Dirac, Werner Heisenberg, and Erwin Schrödinger, 1933. Max-Planck-Institut für Physik, courtesy of AIP Emilio Segrè Visual Archives.
Page 106. Photo by Paul Ehrenfest Jr. Courtesy of AIP Emilio Segrè Visual Archives.
Page 107. Watercolor by Maryke Kammerlingh-Onnes (no date). AIP Emilio Segrè Visual Archives, gift of Stan Fraydas.
Page 108. "A Medicine Show in Huntingdon, Tennessee," October 1935. Photo by Ben Shahn for the Farm Security Administration and Office of War Information, LOC.
Page 109. Photo by Schloss, AIP Emilio Segrè Visual Archives.
Page 111. "Jesse Owens with Lutz Long and Naoto Tajima salute during awards ceremony for the broad-jumping event at the 1936 Olympics in Germany." ACME Newspictures, Inc., LOC Prints and Photographs Collection.
Page 114. "Sir Ernest Rutherford, head, facing slightly left," between 1920 and 1937. LOC Prints and Photographs Collection.
Page 115. "Golden Gate Bridge from the Southwest," September 3, 1937. LOC Prints and Photographs Collection.
Page 116. National Archives and Records Administration, AIP Emilio Segrè Visual Archives.
Page 119. Photo by Alice Calaprice, 1999.
Page 120. Left to right: Margot Einstein, Helen Dukas, Albert Einstein, Maja Einstein, and unknown older woman and boy, World's Fair, 1939. *New York Times.*
Page 122. Photo ca. 1939. National Archives and Records Administration, AIP Emilio Segrè Visual Archives.
Page 123. Churchill radio address, 1939. © Bettmann/Corbis. *New York World-Telegram* and the *Sun Newspaper* Collection, LOC.
Page 124. American Jewish Archives; Brown Brothers; International News Photos, New York; Keystone Press; United Press International; Argonne National Laboratory; AIP Emilio Segrè Visual Archives.
Page 126. "Aerial photograph, taken by a Japanese pilot, of the destruction of Pearl Harbor, Japanese bomber in lower right foreground," published 1956, taken 1941. NEA Services photograph, *New York World-Telegram* and the *Sun Newspaper* Photograph Collection, LOC.
Page 127. Sculpted face of Abraham Lincoln and construction equipment on Mount Rushmore, South Dakota, 1938. Photo by Charles d'Emery of Manugian Studios, South Norwalk, Connecticut; LOC Prints and Photographs Collection.
Page 128. Count Basie's orchestra at the Savoy Ballroom, Chicago, 1941. Farm Security Administration and Office of War Information Collection, LOC.
Page 130. "Stalin, Churchill, and Roosevelt," 1943. U.S. Army/U.S. Signal Corps.
Page 130. Evelyn Chartrand tightening the nose plugs on 500-pound aerial bombs. "Women had a big share in filling these bundles for Berlin in a Canadian plant ...," 1942 or 1943. Farm Security Administration and Office of War Information Collection, LOC.
Page 131. "Negro Marines prepare for action at Camp Lejeune," 1943. U.S. Office of War Information, Overseas Picture Division.
Page 132. Einstein with Capt. Geoffrey Sage and Lt. Cdr. Frederick L. Douthit, both U.S. Navy officers, 1943. U.S. Navy Photo Collection.
Page 133. AIP Emilio Segrè Visual Archives.
Page 134. Photo ca. late 1940s. From the author's collection.
Page 136. "H-bomb cloud over Nagasaki." USAAF, 1945 #A-58450.
Page 137. Einstein, 1945. Copyright not renewed, John D. Schiff, New York, N.Y., LOC Prints and Photographs Collection.
Page 140. Photo ca. 1945. LOC, courtesy of AIP Emilio Segrè Visual Archives.
Page 142. "Chuck Yeager beside Bell X-1." © Bettmann/CORBIS.
Page 142. "The Scroll of the Rule." © The Dead Sea Scrolls Foundation, Inc.
Page 144. Photo ca. 1948. *Sky and Telescope,* AIP Emilio Segrè Visual Archives.
Page 147. "Einstein's 70th birthday at the Institute for Advanced Study," March 14, 1949. Courtesy of the estate of Howard Schrader.
Page 148. Interior of the dome at Mount Wilson Observatory. This view shows the 100-inch reflector telescope and a Cassagrain observing platform, as seen from the west. LOC Prints and Photographs Collection.
Page 150. Photo ca. 1950s. International Communication Agency, U.S. Information Service, AIP Emilio Segrè Visual Archives, Fermi Film and Landé Collections.
Page 152. "Sen. Joseph McCarthy ... seated at desk, speaking to reporters," 1951. Acme Newspictures, Inc., *New York World-Telegram* and the *Sun Newspaper* Photo Collection, LOC.
Page 153. Einstein with Kurt Gödel, ca. 1952. Courtesy of Richard Arens, University of California at Los Angeles.
Page 154. Photo by Erik Gustafson, AIP Emilio Segrè Visual Archives, Margrethe Bohr Collection.
Page 156. "The discoverers of the structure of DNA, James Watson and Francis Crick, with their model of part of a DNA molecule in 1953." © A. Barrington Brown/Photo Researchers, Inc. © 2003 Photo Researchers, Inc. All rights reserved.
Page 157. © Bettmann/CORBIS.
Page 158. LOC Granik Collection.
Page 159. Photo ca. 1950s. National Archives and Records Administration, AIP Emilio Segrè Visual Archives.
Page 161. Photo by Ulli Steltzer, ca. 1950s, AIP Emilio Segrè Visual Archives.
Page 162. McDonald's in Downy, 1955. © Robert Landau/CORBIS.
Page 165. Photo by Alice Calaprice.